ASCE Manuals and Reports on Engineering Practice No. 131

Canal Automation for Irrigation Systems

Prepared by
the Task Committee on Recent Advances in Canal Automation
of the Irrigation Delivery and Drainage Systems Committee
of the Irrigation and Drainage Council
of the Environmental and Water Resources Institute
of the American Society of Civil Engineers

Edited by
Brian Wahlin, Ph.D., P.E., D.WRE
Darell Zimbelman, Ph.D., P.E., D.WRE

ASCE AMERICAN SOCIETY OF CIVIL ENGINEERS

EWRI ENVIRONMENTAL & WATER RESOURCES INSTITUTE

Published by the American Society of Civil Engineers
Reston, Virginia

Library of Congress Cataloging-in-Publication Data

Canal automation / prepared by the Task Committee on Recent Advances in Canal Automation of the Irrigation Delivery and Drainage Systems Committee of the Irrigation and Drainage Council of the Environmental and Water Resources Institute of the American Society of Civil Engineers ; edited by Brian Wahlin, Ph.D., P.E., D.WRE, Darell Zimbelman, Ph.D., P.E., D.WRE.

pages cm. — (ASCE manuals and reports on engineering practice ; no. 131)

Includes bibliographical references and index.

ISBN 978-0-7844-1368-5 (print : alk. paper) — ISBN 978-0-7844-7861-5 (ebook) 1. Canals—Automatic control. I. Wahlin, Brian. II. Zimbelman, Darell D. III. Environmental and Water Resources Institute (U.S.). Task Committee on Recent Advances in Canal Automation.

TC753.C36 2014

627'.135—dc23

2014023572

Published by American Society of Civil Engineers
1801 Alexander Bell Drive
Reston, Virginia, 20191-4382
www.asce.org/bookstore | ascelibrary.org

Errata: Errata, if any, can be found at http://dx.doi.org/10.1061/13685.

ISBN 978-0-7844-1368-5 (paper)
ISBN 978-0-7844-7861-5 (PDF)
Manufactured in the United States of America.

21 20 19 18 17 16 15 14 1 2 3 4 5

Cover credit: Photo by USDA_ARS, Brian Prechtel

MANUALS AND REPORTS ON ENGINEERING PRACTICE

(As developed by the ASCE Technical Procedures Committee, July 1930, and revised March 1935, February 1962, and April 1982)

A manual or report in this series consists of an orderly presentation of facts on a particular subject, supplemented by an analysis of limitations and applications of these facts. It contains information useful to the average engineer in his or her everyday work, rather than findings that may be useful only occasionally or rarely. It is not in any sense a "standard," however; nor is it so elementary or so conclusive as to provide a "rule of thumb" for nonengineers.

Furthermore, material in this series, in distinction from a paper (which expresses only one person's observations or opinions), is the work of a committee or group selected to assemble and express information on a specific topic. As often as practicable the committee is under the direction of one or more of the Technical Divisions and Councils, and the product evolved has been subjected to review by the Executive Committee of the Division or Council. As a step in the process of this review, proposed manuscripts are often brought before the members of the Technical Divisions and Councils for comment, which may serve as the basis for improvement. When published, each work shows the names of the committees by which it was compiled and indicates clearly the several processes through which it has passed in review, so that its merit may be definitely understood.

In February 1962 (and revised in April 1982), the Board of Direction voted to establish a series titled "Manuals and Reports on Engineering Practice," to include the Manuals published and authorized to date, future Manuals of Professional Practice, and Reports on Engineering Practice. All such Manual or Report material of the Society would have been refereed in a manner approved by the Board Committee on Publications and would be bound, with applicable discussion, in books similar to past Manuals. Numbering would be consecutive and would be a continuation of present Manual numbers. In some cases of joint committee reports, bypassing of Journal publications may be authorized.

A list of available Manuals of Practice can be found at http://www.asce.org/bookstore.

CANAL AUTOMATION FOR IRRIGATION SYSTEMS

Editors

Brian Wahlin, Ph.D., P.E., D.WRE, WEST Consultants, Inc. (USA)
Darell Zimbelman, Ph.D., P.E., D.WRE, Water Systems Operations and Management, LLC (USA)

Lead Chapter Authors

Charles Burt, Ph.D., P.E., D.WRE, Irrigation Training and Research Center (USA) – Chapter 1 and Chapter 2
Robert Strand, U.S. Arid-Land Agricultural Research Center (USA) – Chapter 3
Peter-Jules van Overloop, Ph.D., Delft University of Technology (The Netherlands) – Chapter 4 and Chapter 6
Bert Clemmens, Ph.D., P.E., D.WRE, WEST Consultants, Inc. (USA) – Chapter 5 and Chapter 7
Sumith Choy, B. Tech, MEngSc., Rubicon Water (Australia) – Chapter 8

Contributing Authors

Gilles Belaud, Montpellier SupAgro (France)
Jean-Luc Deltour, Société du Canal de Provence (France)
Pierre-Olivier Malaterre, Irstea (France)
Tony Oakes, Rubicon Water (Australia)
Bob Gooch, Salt River Project (USA)
Irrigation Training and Research Center Staff (USA)

BLUE RIBBON REVIEW PANEL

Robert Norman, U.S. Bureau of Reclamation (USA)
Gary Merkley, Consultant (USA)
David Rogers, Rogers Engineering Hydraulics, Inc. (USA)
Stuart Styles, Irrigation Training and Research Center (USA)

CONTENTS

PREFACE

Canal automation always has had the potential to save water and improve efficiency of irrigation water supply projects or of irrigation district operations. Recently, there have been a number of technological and engineering advances in the field of canal automation. The Task Committee on Recent Advances in Canal Automation was formed under the Irrigation Delivery and Drainage Systems Committee under the Environmental and Water Resources Institute of the American Society of Civil Engineers to document the new technological progress in canal automation. Members of the task committee gathered information on canal automation research that is taking place around the world. The task committee was truly an international effort with researchers and engineers in several countries (USA, The Netherlands, Australia, France, Spain, Portugal, China, and Mexico) all participating in the development of this Manual of Practice. This publication is designed to provide guidance on how and when to implement canal automation within the context of canal modernization but not covering the full range of canal modernization issues. The manual also provides practical guidance on some of the more routine aspects of canal automation.

DEVELOPMENT OF THE MANUSCRIPT

The task committee kicked off the project with a series of three short video conferences October 13–15, 2009. The video conference was hosted at three sites across the world (USA, Australia, and France). This video conference was unique in that it allowed a large group of canal automation experts to come together to discuss the manual of practice without undue travel. During the video conference, participants discussed the overall goals of the Canal Automation Manual, developed an outline of the manual, and made initial assignments for writing the chapters. The task committee would like to thank the hosts of the three video conference sites:

- Arid Land Agricultural Research Center (USA),
- University of Melbourne (Australia), and
- Montpellier SupAgro (France).

Participants in this initial video conference included

- David Aughton, Rubicon Water (Australia),
- Jean-Pierre Baume, Cemagref (France),
- Eduardo Bautista, U.S. Arid-Land Agricultural Research Center (USA),
- Gilles Belaud, Montpellier SupAgro (France),
- Charles Burt, Irrigation Training and Research Center (USA),
- Sumith Choy, Rubicon Water (Australia),
- Bert Clemmens, WEST Consultants, Inc. (USA),
- Jean-Luc Deltour, Société du Canal de Provence (France),
- Tugba Erol, Rubicon Water (Australia),
- Guy Fipps, Texas A&M University (USA),
- Olivier Genthon, Société du Canal de Provence (France),
- Tom Gill, U.S. Bureau of Reclamation (USA),
- Manuel Gomez, Technical University of Catalunya-Barcelona (Spain),
- Bob Gooch, Salt River Project (USA),
- Guan Guanghua, Wuhan University (China),
- Remy Halm, Rubicon Water (USA),
- Klaudia Horváth, Technical University of Catalunya-Barcelona (Spain),
- Xavier Litrico, Lyonnaise des Eaux à Bordeaux (France),
- Pierre-Olivier Malaterre, Irstea (France),
- Juan Mantecon, Technical University of Catalunya-Barcelona (Spain),
- Iven Mareels, University of Melbourne (Australia),
- Luciano Mateos, University of Cordoba (Spain),
- Tony Oakes, Rubicon Water (Australia),
- Manuel Rijo, University of Evora (Portugal),
- Jose Rodellar, Technical University of Catalunya-Barcelona (Spain),
- Carlos Sepulveda, Technical University of Catalunya-Barcelona (Spain),
- Robert Strand, U.S. Arid-Land Agricultural Research Center (USA),
- Blair Stringam, New Mexico State University (USA),
- Peter-Jules van Overloop, Delft University of Technology (The Netherlands),

- Yann Viala, Société du Canal de Provence (France),
- Brian Wahlin, WEST Consultants, Inc. (USA),
- Min Xu, Deltares, (The Netherlands), and
- Darell Zimbelman, Water Systems Operations and Management, LLC (USA).

TASK COMMITTEE WORKSHOP

From September 17 to 20, 2012, the Task Committee on Canal Automation held a workshop at the Société du Canal de Provence in Aix-en-Provence, France. The goal of the workshop was to have a face-to-face meeting to discuss the Canal Automation Manual, as well as to share ideas and concepts. The Task Committee would like to thank the Société du Canal de Provence for organizing and hosting the event. Participants included

- Jean-Luc Deltour, Société du Canal de Provence (France),
- Bruno Grawitz, Société du Canal de Provence (France),
- Bruno Vergobbi, Société du Canal de Provence (France),
- Pierre-Olivier Malaterre, Irstea (France),
- Gilles Belaud, Montpellier SupAgro (France),
- Darell Zimbelman, Water Systems Operations and Management, LLC (USA),
- Charles Burt, Irrigation Training and Research Center (USA),
- Bert Clemmens, WEST Consultants, Inc. (USA),
- Robert Strand, U.S. Arid-Land Agricultural Research Center (USA),
- Peter-Jules van Overloop, Delft University of Technology (The Netherlands),
- Sumith Choy, Rubicon Water (Australia),
- Manuel Gomez, Technical University of Catalunya-Barcelona (Spain),
- Klaudia Horváth, Technical University of Catalunya-Barcelona (Spain), and
- Luciano Mateos, University of Cordoba (Spain).

ACKNOWLEDGMENTS

The task committee is deeply indebted to Monica Holman at the Irrigation Training and Research Center. Ms. Holman spent countless hours editing and formatting the manual. Her hard work and dedication made this manual a much better document. The task committee also would like to thank the Irrigation Training and Research Center for allowing Ms. Holman to work on the manual.

Unless otherwise noted, all pictures and figures were provided by the lead chapter author. The editors of this manual would like to express their appreciation to all the lead authors for their dedication to this long-term project.

CHAPTER 1

MODERNIZATION PROCESS, CONSTRAINTS, AND CONCEPTS

1.1. INTRODUCTION

This chapter provides a general overview of the potential benefits of irrigation canal modernization and discusses considerations for assessing whether an irrigation or water district should implement automation as it modernizes, and how to proceed.

1.2. MODERNIZATION

Modernization is a combination of technical, managerial, and organizational upgrading (as opposed to mere physical rehabilitation) of irrigation schemes with the objective of improving resource utilization (e.g., water, labor, economics, or environment) and water delivery service to farms (Wolter and Burt 1997). Such modernization investment focuses on the details of the *inner workings* of an irrigation project. Planners and engineers for irrigation projects frequently equate modernization with practices such as canal lining, piping, and computerized automation; however, such investments are often of low initial priority if one examines the steps needed to improve performance. Computerized automation typically is implemented in later stages of modernization, after basic needs such as flow measurement and accounting procedures have already been completed. Modernization is a *process* that sets specific objectives and selects specific actions and tools to achieve them over an extended period.

Modernization that is intended to respond to external pressures (such as improved return flow quality or reduced diversions) simultaneously should address improved service to the farm/field/customer. Irrigation systems must serve the farmers well enough that they attain high on-farm production and irrigation efficiencies and can afford to pay the water service fee. The farmers are, after all, the eventual recipients of the irrigation project's outcomes, and it is important to remember that the primary purpose of the project is to support agriculture. Additionally, the fees paid by farmers are often the source of funding for system improvements. Although the focus of this publication is systems delivering water to farmers for irrigation, the contents apply equally well to canal systems delivering water to a variety of destinations, including potable water suppliers.

The focus of modernization efforts (i.e., technical, managerial, or organizational) has varied with time and location. The World Bank, for example, has shifted its focus several times over the past few decades. In the 1980s the emphasis of training was on upgrading a combination of all aspects of the irrigation system, although irrigation project contracts contained little, if any, language regarding modernization or expected hydraulic and service requirements. In the late 1980s and early 1990s, the emphasis shifted to physical technical interventions such as improved water level and flow control. In the early 2000s the emphasis shifted back almost completely to the creation of water user associations. This emphasis on *social institution* modernization continues today. However, the success of a social institution that is intended to control and distribute water is highly dependent on whether the water supply is manageable in the first place. In other words, having the proper engineering design that provides excellent hydraulic control is essential for achieving social goals such as equity and transparency.

This manual of practice focuses on the technical aspects of controlling water on a minute-to-minute basis throughout parts or all of an irrigation project. This is not to say that the managerial, organizational, and general infrastructure upgrading is any less important. However, many social problems in areas with severely limited water supplies can be eliminated when water is controlled easily and supplied to customers, users, or operators farther downstream in the hydraulic network with a high degree of service (i.e., with flexibility, reliability, and equity). Conversely, a water supply that cannot be managed easily and efficiently rarely can accomplish the economic and social goals that were predicted during the project's concept stage.

In the United States, France, Australia, and many other countries, there are often strong and effective local water user associations. Irrigation projects with a strong water user association have effective bylaws and enforcement regarding water delivery equity, theft, and vandalism. Therefore, the application of control systems that provide high flexibility has received the most attention; social issues remain but are not nearly as difficult as in some other countries.

The task today is largely one of modernizing existing branching networks, or "layers," as opposed to designing new irrigation projects. From a technical control standpoint, which is the focus of this publication, modernization experts typically view an irrigation system as a series of layers through which water passes. Each layer has an obligation and needs incentives to provide a good level of service to the next lower layer (Burt and Styles 1999).

There are no single answers as to how and when to implement automation as part of modernization because of the complexity and variety of combinations of the water supplies (surface versus conjunctive use), water quality, timing of flows, adequacy of the water supplies, topography, aquatic weed problems, soil types as related to seepage and bank stability, usage of return flows, types of existing structures, and so on. There are, however, some basic principles, as presented here, that should be followed to achieve a high level of success in designing, constructing, and implementing a canal automation project. An important consideration is always whether and how existing water control structures can be incorporated into a modernization scheme, as opposed to replacing all existing water control structures.

Specifically, this publication focuses on the technical aspect of upgrading canal operations through the use of an automated control system. It is recognized that this is only one aspect of modernization. There are numerous other physical aspects of irrigation project modernization, such as proper sizing of canals for flexibility, canal lining, having road access on both sides of a canal for maintenance, incorporation of pipelines, and flow measurement devices. Further, excellent water level control often can be achieved by simple structures such as long-crested weirs, flap gates, or other hydraulic gates, without the need for electronic controls or power at the site. Because of its importance and complexity, this publication focuses on automatic control, including computerized control.

1.3. POTENTIAL BENEFITS OF MODERNIZATION AND AUTOMATION

Small engineering, design, or management changes often can be made here or there with noticeable positive effect. For example, new tools such as handheld data recorders can scan bar codes on turnouts (sometimes called delivery gates or offtakes) to upload field observations of flow rate and/or volume or to manually record crops grown and delivery starts and stops. The information can be downloaded at the office and automatically entered into billing software or linked directly with radio communications. Another example of a simple improvement is to use existing side boards (overflow weirs) on a water control sluice gate structure to achieve a weir effect and provide much better water level control. Such aspects of modernization are relatively inexpensive, both in terms of initial investment and operations and maintenance (O&M).

The cost of a major modernization project, however, can be high. Major modernization on a large project can cost tens of millions of U.S. dollars. A single structure may cost more than $100,000. There must, therefore, be benefits associated with modernization and automation to justify the investment.

The real question, of course, might be, "Can the world afford *not* to modernize irrigation?" In brief, the facts are these:

1. The increased yields of grain since 1970, worldwide, are largely attributed to new varieties (the "green revolution") and expanded irrigated acreage.
2. There are no spectacular new high-yield grain varieties expected in the near future.
3. It is now recognized that further expanding irrigated acreage generally is not feasible, as many water supplies are already overallocated. This is evidenced by widespread groundwater overdraft and environmental degradation of rivers and lakes in many locations.
4. The world's population continues to increase, and in many countries people are now consuming more meat, which requires extra crop production.
5. Irrigated agriculture is expected to produce most of the additional food needed to feed the increasing population.

These points clearly indicate that there must be an improvement of crop production per unit of water consumed. It is now generally understood that there must be simultaneous protection of the environment. It is only with improved water control and management that these objectives can occur. Hence, the true issue is *how* modernization (possibly including automatic control), should proceed, rather than *whether* modernization should proceed at all.

Nevertheless, this global view of modernization does not pay the immediate bills related to modernization investment. Local irrigation agencies typically must identify specific benefits to justify the expense and effort of modernization. Fortunately, the benefits are numerous. Of course, any discussion of improvement assumes that one knows the present status of a project and has an effective strategy for improvement. Methods for achieving this are discussed briefly later in this chapter. Typical benefits that are observed are discussed as follows, including

1. Supply-oriented versus demand-oriented operations;
2. Improved water deliveries for maximizing crop production;

3. Adoption of modern on-farm irrigation systems;
4. Improved river (in-stream) flows and river quality;
5. Reduced spills and tail-end losses;
6. Increased efficiency and accountability;
7. Elimination of operator "kingdoms";
8. Improved social harmony;
9. Reduction of labor, lawsuits, and energy costs;
10. Increased flow rate capacity; and
11. Compliance with environmental law and water rights.

1.3.1. Supply-Oriented versus Demand-Oriented Operations

A common problem with the operation of large water delivery systems is that they are often supply oriented. When the water flow rate changes in a large canal, a sudden change upstream arrives gradually downstream because of wave dispersion. More discussion of this topic will be presented in later chapters, but simply put, this can make operation of a large canal with frequent changes difficult. A supply orientation favors rotational water delivery schedules, where the supply released at the upstream end does not change, and water is rotated among smaller canals and eventually users. Such a system almost eliminates the ability of farmers to optimize on-farm irrigation performance, and it does not allow farmers to adapt to market needs by changing crops.

When there is a desire to be more responsive to farmers and at the same time to be more efficient with the delivered water, improved operational methods and controls typically are required for a canal system. Modernization that includes a demand orientation is the overall focus of this publication. Accomplishing all potential improvements also may require changes or the adaptation of new policies by the managing authorities; such policy changes are very project-specific and beyond the scope of this publication.

1.3.2. Improved Water Deliveries for Maximizing Crop Production

There is no doubt that the high crop yields that are needed in the near future can be obtained only if there is excellent on-farm irrigation management. The irrigation of fields that are supplied with canal water on a rotation schedule, as an example, cannot be managed to achieve a high degree of agricultural productivity. To achieve simultaneous high levels of productivity and irrigation efficiency, the water must be delivered with excellent service, considering aspects of flexibility, equity, and reliability. Unless the canal system is properly controlled with the required hardware and software (which includes human input as well as possible electronic controls), the water delivery service will suffer unless there is significant water that spills at the tail ends of the canal system.

The quality of water delivery service also influences drainage problems. Water that cannot be applied to a field on a timely basis, and with the correct volume, can cause serious surface and subsurface drainage problems. This, in turn, makes removing salt from the root zone difficult. The combined problems of high salinity, high water tables, and fluctuating water tables all contribute to suboptimal crop yields.

1.3.3. Adoption of Modern On-Farm Irrigation Systems

Farmers in most areas do not purchase modern on-farm irrigation systems to conserve water; they purchase them to improve crop yields or quality or to grow a new crop type. Drip and sprinkler systems of course need flexible and reliable water deliveries; they cannot be supplied with rotation deliveries unless on-farm storage ponds are constructed at the expense of productive acreage. For example, irrigation districts quickly learn that providing flexible deliveries, of low flow rate per field, to drip systems is quite challenging because the flows in the canals constantly fluctuate. This is especially challenging if farmers use electric pumps and want to take advantage of lower power rates by not pumping during high-cost periods of peak electrical demand.

1.3.4. Improved River (In-Stream) Flows and River Quality

There are a number of old irrigation projects that offer excellent water delivery service to farmers on a demand basis. These old projects have almost no management or complex structures. They rely on a simple principle: divert very large flow rates at the head of the canal and allow most of that water to reenter the river downstream of the project. This means that water is available at the turnouts on demand. This was clearly a simple and inexpensive means of providing good service, and downstream users still obtained their water, but it came at a cost to in-stream flow rates and the associated aquatic habitat. Now, as many endangered species dependent on the river are targeted for restoration, these old, simple projects must be almost completely modernized to achieve the simultaneous goals of diverting much less water (to keep the river flowing downstream of the diversions) and continuing to provide the same degree of water delivery flexibility.

The flow rate in a river below an irrigation diversion often also influences the river's water quality. On a localized basis, lower in-stream flow rates often are incapable of sufficiently blending lower quality irrigation return flows to prevent adverse effects to water quality. Better, more efficient, first-time usage of diverted water can decrease the amount of water diverted from the river, resulting in more in-stream flow adjacent to the project and improved river water quality downstream.

Improved water control within an irrigation district or project often is needed to be able to recycle and blend drainage flows within the district to avoid discharging those drainage flows back into the rivers. In some projects, improved surface water management allows a project to stop pumping lower quality well water.

1.3.5. Reduced Spills and Tail-End Losses

Because modernization emphasizes better real-time information and technology to provide improved water level and flow control, canal losses (other than seepage) usually can be reduced significantly. Complete elimination of tail-end losses, while still providing excellent water delivery service, is very difficult and expensive with large systems. Some modernization schemes incorporate spill recapture systems that recycle such losses, rather than attempting to eliminate them.

1.3.6. Increased Efficiency and Accountability

Some irrigation projects may be facing reductions in allowable diversions. Modernization may be essential to minimize conveyance and distribution losses so that with the reduced diversion, the same volume of water is still available to the farmers, it is hoped, with even better water delivery service.

If an irrigation project lacks the proper measurement and control structures, it likely has a high level of management uncertainty and very poor accounting. The operators do not know exactly where water went or when. A manager who drives out to a canal has no way of knowing (other than from farmer complaints) if a canal operator is operating the system properly. There are always excuses from operators about problems being caused by upstream operators, water lag time, or lack of farmer discipline. It is almost impossible for a manager to be able to decipher the situation in the field quickly at any particular moment. The system is run as an "art," which depends on nontransferable knowledge. Modernization helps to eliminate the art of water delivery.

If flow rates into canals or turnouts fluctuate uncontrollably with time, operators and farmers always have an excuse for poor management on their part. Operators must somehow compensate, often by not recording deliveries properly or by underestimating delivery flow rates (assuming that some time during the delivery, the flow may drop). Modernization, if done properly, provides the control and measurement that enables everyone to be readily accountable for water received and delivered. The irrigation system operation becomes a professionally operated process.

1.3.7. Elimination of Operator Kingdoms

This is directly related to improved professionalism and accountability. In many projects, the operators work out special deals with farmers. They treat some farmers better than others. Further, the operation of their canals is so complex that a new operator needs months if not years to learn how to operate the system. Consequently, a manager is reluctant to eliminate the special deals or problems with a particular operator by replacing the operator and training a new person. An easily understood and manageable canal system helps to minimize or eliminate these problems. Following the implementation of automatic control, it is relatively easy to train a new operator, and it is also easy to verify the performance of existing operators.

1.3.8. Improved Social Harmony

Water that is well-controlled, flexible, and accountable enables the operators to have a much smoother relationship with their customers. If farmers clearly understand their entitlement and are assured of a flexible and well-measured supply, typically they are satisfied with their water delivery service and have few complaints. Farmers who perceive that they are receiving equal treatment are also less likely to develop animosity or to file complaints with their district management or with their neighbors.

1.3.9. Reduced Costs

Reduction in *labor* costs is not necessarily a given with canal automation. Modern electronics require constant maintenance. Further, supervisory control and data acquisition (SCADA) systems may require additional personnel looking at data 24 hours per day, 7 days per week. So although field operators may see reduced travel time and vehicle expenses from less frequent visits to the ends or heads of lateral canals to check on spill or to reduce incoming flows, additional people with new skills are needed. Modernization often requires the employment of technicians who can maintain and troubleshoot the automatic control systems, the cost of which must be included in the feasibility estimates at the beginning of the planning process. Employees with the skills and education needed to deal with sensors, communications equipment, and electronics gear demand higher salaries.

Therefore, it is unusual to reduce overall labor costs with a modernization program. This, however, should be considered in another light. With the existing infrastructure in most irrigation projects, the only way to improve flexibility and reliability is to hire many more people (e.g., night operators). With improved infrastructure and information management, the same number of people, some with different skills, can provide much better service.

Some irrigation projects have frequent canal breaks or canal bank overtopping, which often result in lawsuits to recover the cost of reparations on damaged property or injuries. Good water level control also can minimize rodent damage and washouts by managing or eliminating burrowing below the maximum water level during times when the water levels are low. This assumes, of course, that the equipment installed as part of the modernization is operational at all times by being robust and dependable.

Finally, if a project is supplied by a pumped diversion, or from groundwater wells, an improvement in the irrigation efficiency will result in less gross water needed. This, in turn, means that well pumps may be shut off more frequently or the diversion pumping may be less. Modernization projects that involve lift stations often incorporate variable frequency drive (VFD) controllers into the control logic, which eliminates the need for physical recirculation of water around the pumps to achieve the desired level of service. Conserving energy through reduced pumping has many project and nonproject benefits.

1.3.10. Increased Flow Rate Capacity

If water levels can be controlled safely within a tight range, less freeboard is needed. This, in turn, allows operators to pass more water safely (with a higher water surface) through a canal than previously. This increased operational capacity can be used to satisfy high demands or to improve the level of service.

1.3.11. Compliance with Environmental Law and Water Rights

The immediate stimulus for modernization is often external pressures, including lawsuits regarding water rights or new regulations on diversions and river water quantity or quality. Modernization typically is required to respond to the new rules. The biggest challenge to engineers is to design a modernization program that not only satisfies the external pressures but that also simultaneously improves water delivery service to farmers, all at a reasonable cost.

1.4. ASSESSMENT OF AN EXISTING IRRIGATION SYSTEM PRIOR TO MODERNIZATION

A specific procedure to evaluate and benchmark irrigation project delivery systems was developed by Burt and Styles (1999) for the World Bank and was subsequently adopted by UN/FAO. It was named the rapid appraisal process (RAP) and was the first procedure of its type to provide a comprehensive evaluation of the internal and external processes associated with irrigation water delivery. To make the procedure systematic, detailed data and computational Excel spreadsheets are available for evaluators to use. This procedure has been extended by FAO in the MASSCOTE (Renault et al. 2007) process to include additional concepts such as mathematical sensitivity of structures and flow capacity analysis mapping.

An essential ingredient of the successful application of these RAPs (and therefore, its greatest weakness) is adequate training and expertise of the evaluators. Experience has shown that successful RAP programs require evaluators with prior training in irrigation, specific training in the RAP techniques, and follow-up support and critique when the evaluators begin their fieldwork. The importance of having qualified evaluators cannot be overemphasized.

Typical baseline data for external indicators (such as water balances and irrigation efficiency) are either readily available or they are not available at all. Individual irrigation projects have differences in the ease of access to typical baseline data on the irrigation project, weather, and water supply, among other things. In some projects the data can be gathered in a day; in others, collection may take weeks.

A quick and focused examination of irrigation projects can give a reasonably accurate and pragmatic description of the status of the project and the processes and hardware that influence that status. This allows for the identification of the major actions that can be taken quickly to improve water delivery service, especially if the RAP is conducted in cooperation with the local irrigation authorities.

The question of what is "reasonably accurate" in data collection and computations always can be debated. Confidence intervals should be assigned to most water balance data, reflecting the reality that there are always uncertainties in data and computation techniques. In irrigation matters in general, 5 to 10% accuracy is a typical range; better accuracy often is not achievable in practice (Clemmens and Burt 1997). The problems one encounters in irrigation projects are typically so gross and obvious (to the properly trained eye) that it is unnecessary and inappropriate to strive for extreme accuracy when one wants to diagnose their functionality.

1.4.1. External Indicators

External indicators for irrigation projects are ratios or percentages that generally have forms such as

$$\frac{\text{Water Required}}{\text{Total Water Available}} \quad \text{or} \quad \frac{\text{Crop Yield}}{\text{Irrigation Water Delivered to the Fields}}$$

The common attribute of external indicators is that they examine inputs and outputs for a project. External indicators are expressions of various forms of efficiency, whether the efficiency is related to budgets, water, or crop yields. Even more than that, they only require knowledge of inputs and outputs to the project. By themselves, external indicators do not provide any insight into what must be done to improve performance or efficiency. The identification of what actions must be taken to improve these external indicators comes from an analysis of internal indicators, which examine the processes and hardware used within the project.

However, external indicators do establish key values, such as whether it might be possible to conserve water (without defining how that might be accomplished). As such, low values of external indicators often provide the justification for modernization of projects, with the anticipation that modernization or intervention will improve the values of those external indicators.

The external indicators focus on items of a typical large-scale water balance. Values such as crop evapotranspiration, effective precipitation, and water supplies must be estimated. The concept of a water balance, on any scale, is critical to understanding and assessing water management issues.

1.4.2. Confidence Intervals

A certain amount of error or uncertainty is inherent in all measurement or estimation processes. Therefore, it is impossible to know the true or correct values for the water volumes needed to calculate terms such as "irrigation efficiency." Estimates must be made of the component volumes based on measurements or calculations.

In reports that provide estimates of crop yield and water balance ratios with terms such as "irrigation efficiency" and "relative water supply," the uncertainties associated with those estimates should be acknowledged and quantified. Otherwise, planners may not know if the true value of a stated 70% efficiency lies between 65% and 75% or between 50% and 90%.

One method of expressing the uncertainty in a single-valued estimate is to specify the confidence interval (CI) for that estimate. If it is believed that a reasonable evaluation of data indicates that the correct value lies within 5 units of 70, then it should be stated that the quantity equals 70±5. More specifically, the essence of a confidence interval should be illustrated as follows when discussing an estimated quantity: The investigators are 95% confident that their estimate of the irrigated area in the project is within ±7% of 500,000 ha (between 465,000 ha and 535,000 ha).

1.4.3. Internal Processes and Internal Indicators

Broad goals of modernization are to achieve improved irrigation efficiency (an external indicator), better crop yields, less canal damage from uncontrolled water levels, more efficient labor, improved social harmony, and an improved environment as accomplished by fewer diversions or better quality return flows. In general, these goals can be achieved only by paying attention to *internal* details.

As an example of the usage of internal indicators used in the original RAP (Burt and Styles 1999), primary indicator I-1 is used to characterize the actual water delivery service to individual ownership units. Primary indicator I-1 has four sub-indicators:

I-1A. Measurement of volumes to the field
I-1B. Flexibility to the field
I-1C. Reliability to the field
I-1D. Apparent equity.

The value for each primary indicator (e.g., No. I-1) can be computed automatically in an "internal indicators" worksheet by

1. Applying a relative weighting factor to each sub-indicator value; the weighting factors are only relative to each other within the indicator group; one group may have a maximum value of 4, whereas another group may have a maximum value of 2; the only factor of importance is the relative weighting factors of the sub-indicators within a group,
2. Summing the weighted sub-indicator values,
3. Adjusting the final value based on a possible scale of 0 to 4 (4 indicating the most positive conditions).

Each sub-indicator (e.g., No. I-1A) has a maximum potential value of 4.0 (best), and a minimum possible value of 0.0 (worst). Each sub-indicator score also is weighted depending on a predefined importance (weight). The sum of the weighted sub-indicator values equals the primary indicator value. Table 1-1 lists classifications of primary indicators from Burt and Styles (1999).

1.5. TYPICAL MODERNIZATION STEPS

As a general guideline, the following steps are recommended for the modernization process:

1. *Define a current level of service to users.* Eliminate the discrepancy between "actual" and "stated" service. Project management must be

Table 1-1. Information for Primary Indicator I-: Actual Water Delivery Service to Individual Ownership Units (e.g., Field or Farm)

No.	Sub-Indicator	Ranking Criteria	Wt
I-1A	Measurement of volumes to the individual units (0–4)	4—Excellent measurement and control devices when properly operated and recorded 3—Reasonable measurement and control devices with average operation 2—Useful but poor measurement of volumes and flow rates 1—Reasonable measurement of flow rates but not of volumes 0—No measurement of volumes or flows	1
I-1B	Flexibility to the individual units (0–4)	4—Unlimited frequency, rate, and duration but arranged by users within a few days 3—Fixed frequency, rate, or duration but arranged 2—Dictated rotation but approximately matches the crop needs 1—Rotation deliveries but on a somewhat uncertain schedule 0—No established rules	2
I-1C	Reliability to the individual units (0–4)	4—Water always arrives with the frequency, rate, and duration promised; volume known 3—Very reliable in rate and duration with occasionally a few days of delay; volume known 2—Water arrives about when needed and in the correct amounts; volume unknown 1—Volume unknown; deliveries fairly unreliable but less than 50% of the time 0—Unreliable frequency, rate, duration more than 50% of the time; volume unknown	4

Table 1-1. Continued

No.	Sub-Indicator	Ranking Criteria	Wt
I-1D	Apparent equity to individual units (0–4)	4—All fields throughout the project and within tertiary units receive the same type of water delivery service 3—Areas of the project receive the same amounts of water but within an area the service somewhat inequitable 2—Areas of the project receive somewhat different amounts (unintentionally) but within an area is equitable 1—Medium inequities both between areas and within areas 0—Differences of more than 50% throughout the project on a fairly widespread basis	4

at least willing to develop a better understanding of their system if improved operational control is desired. If project management does not accept reality, the system will not improve.

2. *Understand and adopt a service mentality with all project staff.* This must apply to all levels of staff. Of course, this is not done overnight, but modernization concepts are rooted in this mentality. Without it, attempts to modernize a project typically have minimal benefit.
3. *Train staff in service-oriented operations.* Closely examine instructions that are given to operators, and modify them if they do not meet the proper operational goals. A classic example is found in many projects in which the objective of cross regulators is to maintain an upstream water level, but the gate operators must move the cross regulators in strict accordance with instructions (of specific gate movements) from the office, based on computer programs or spreadsheets. A simple check in the field will show that water levels are not maintained properly. The type of instructions given to the operators must be changed, and they are simple: "Maintain the upstream water level within a specified tolerance of a defined target." It would be extremely rare to find an operator who is incapable of determining how much to move the cross regulator gates to achieve this goal.

These first three items appear simple, yet they may also be the most difficult to accomplish. If the three items cannot be achieved, little progress will be made. Changes in these three items may take some training, study tours, deep conversations, time, and reflection by senior management.

The next steps, more or less in order of sequence, are to improve the following areas:

4. *Understand what conditions lead to poor delivery service for this project.* Experts can evaluate a project quickly and, because of their background, almost immediately understand cause and effect relationships and the probable level of service. Operators and supervisors seldom see things the same way. It is very helpful to install simple data loggers and water level sensors at key locations to record spills, flow rate fluctuations, and water level fluctuations. This is almost always an eye-opener for operators who can only visit a location once per day and who in many cases do not make measurements.
5. *Select an appropriate canal operation strategy.* This is discussed in more detail in a later section.
6. *Improve communications at all levels.* This starts with human-to-human communications, often with radios and cell phones.
7. *Improve staff mobility.* In general, a small yet mobile staff is much more efficient than a large, relatively immobile staff. This is because a small, mobile staff is not responsible for only one or two structures per person; an individual must understand how various structures and actions affect other areas. Mobility may be improved with better roads, motorcycles, and trucks. A SCADA system is an alternative for providing the benefits of staff mobility as it allows operators to see what is happening and move gates without actually traveling to the site.
8. *Provide excellent flow rate control and measurement at key bifurcation points.* Note that "measurement" and "control" are not the same. Both are needed. There are many combinations of structures and techniques that provide rapid and accurate control and/or measurement of flow rates. The flow rate control and measurement are essential at the heads of canals and pipelines and at delivery points. This is typically a weak area for irrigation projects; it hinders quantitative assessments of water management, and it makes it difficult to implement an operations plan when flow rates and volumes are unknown.
9. *Construct recirculation points and/or buffer (balancing) reservoirs in the main canal system.* Because of the long wave travel times, uncertainties with flow measurement and control, and the flexibility of

deliveries in a modern system, buffer reservoirs are very popular to achieve physical points of re-regulation within canal systems. (This is not always a viable option.)

10. *Improve water level control throughout the project.* If flow rate is controlled at the canal heading, the goal at structures downstream is to maintain fairly constant water levels so that gravity-turnout flow rates do not change with time and so that the canal banks are not damaged. With the proper types of structures, this is easy to do without much effort.
11. *Reorganize procedures for ordering and dispersing water.* In most modern projects, one group is responsible for operating the main canal, another is responsible for the second level, and so on. Each group, then, has a specific service objective. If a main canal is broken into "zones" with different offices controlling different zones, there is almost always conflict between the zones. Reorganization of the operators typically is necessary to achieve a true service mentality. The procedure for receiving farmer demand for water from the field and responding to requests typically must be revamped for most projects if a true service mentality is desired. In the western United States, many modernized districts use commercial software to assist in centralized daily organization of water orders that are requested by farmers via the Web or by phoning into a central office.
12. *Provide remote manual control of flow rates at strategic locations.* Locations are at the heads of the main canal and heads of major offtakes (turnouts) from the main canal. This generally requires some form of SCADA system, along with programmable logic controllers (PLCs) or remote terminal units (RTUs) at the field sites. (More detail is provided in later chapters.)
13. *Provide for capture and recording of spills.* Locations are buffer reservoirs and the tail (downstream) ends of canals.

What may surprise some readers is the complete lack of discussion of canal lining and maintenance equipment. There is no doubt that maintenance equipment must be adequate. Canal lining also can reduce maintenance and seepage and provide important bank stability. Lining and maintenance have been discussed for many decades, but the vast sums that have been spent on canal lining generally have not brought about significantly improved water delivery service to users, which is clearly needed to satisfy the looming food shortages. In fact, concrete lining of canals (often because it is poorly designed and installed) is typically followed by eventual deterioration and the need to line again—a vicious cycle of rehabilitation and deterioration. Items 4 through 13 represent a departure from traditional thinking of "concrete" civil engineers and place the focus on operations.

The 13 items listed presume that the project started without automation and used conventional upstream control. It is often most beneficial to adopt all or some of the 13 items before attempting systemwide automation. Gradual improvement of upstream control provides the project staff with a relatively risk-free process of becoming familiar with the technology associated with modern control, such as radios, water level sensors, PLCs, SCADA systems, and other things.

The chapters in this manual of practice focus on more sophisticated PLC canal control, which enables irrigation projects to achieve higher levels of performance and service than with strictly manual systems. However, there is absolutely no doubt that PLC-based control methods require significant annual budgets and the technical skills necessary to operate and maintain this type of electronic equipment. Modern but manual upstream control with some sophisticated technology allows a project to "walk before it runs," taking a step-by-step, incremental approach.

1.6. SELECTING THE APPROPRIATE CANAL OPERATION STRATEGY

A "canal operation strategy" in this discussion refers to a plan for how water will be provided to water users and how depths and flow rates will be controlled on a minute-to-minute basis throughout a project. This includes two important aspects of water control:

1. *Water delivery mode* (determination of the timing and amount of water deliveries to users). Rigid systems determine a schedule far in advance (i.e., at the beginning of the season), whereas modern systems allow changes in farmer water orders sometimes on demand. Because of travel time delays in canals and to ensure that demand does not exceed system capacity, most districts require a lead time for new water orders. The ability of the district to accommodate changes (e.g., early or late shutoff or flow rate changes) depends on the physical layout of the canal system, capabilities of the control system, and policies of the district.
2. *Control strategy*. A designer must not only identify where and why specific controls will be located but also understand the details of how the design influences the manageability and the ability to provide delivery service. This also includes understanding how actions at each point influence actions and results at other locations in the system.

The classical language found in literature usually includes "upstream control" and "downstream control." These generally apply only to water level control. The process of releasing water ahead of time so that it will be available downstream to supply water to a user is often called "routing" or "feedforward control." These and other control options are discussed later in this publication.

In the early days of canal automation, designers viewed the canal system as a complex, branching system. As thinking matured, many successful designers adopted a more simplistic description. In the more simplistic view, an upstream controlled canal has flow rate control at the head of each main and lateral canal and flow rate control at each turnout, with constant upstream water level control throughout the canals. The primary objectives of better control are to control the flow rates carefully at each delivery point (also called offtake or turnout). Further, it is expected that the downstream offtakes would use the same total flow rate as was discharged at the head of a canal. Automation consists of water level control within the conveyance and distribution canals and sometimes automated flow rate control at delivery points. These systems always suffer from the "tail-end problem" of feast or famine because it is impossible to deliver exactly the needed flow rate into a canal and still maintain fairly constant water levels without having flow rate discrepancies at the tail end of the canal.

Downstream control was introduced as a completely automatic method of controlling water levels in canals. Each cross regulator would only deliver the needed flow rate to maintain a target downstream water level. The concept was that there would be no tail-end problem, and communications could be reduced, because the control system would supply more or less water automatically, depending on the downstream withdrawals. Simple hydraulic gates were used at first, but newer modernization schemes sometimes implement computerized automatic gates of various types. These systems are complex and in some sense riskier than upstream control because they actually control the inflow into the canal. Thus, a component failure could cause a canal headgate to open or close completely unless sufficient safeguards are included. This is one reason why early attempts at automation by a district should avoid downstream control. Once a district is comfortable with automatic control, it may find that the added benefits of downstream control will overcome the risks.

In more recent years, targeted recirculation of drainage flows and spills within a project has been found to simplify the control in some systems greatly. Rather than attempt to match the supply and demand exactly everywhere throughout a project (which can be a very complex task), the flows can match approximately what is needed in the upper portions of the project. Spills then are collected downstream, and flow discrepancies in the canals are re-regulated in buffer (regulating or balancing) reservoirs near the downstream ends of canal systems. The control becomes simpler and is more foolproof than downstream control. The monitoring of water levels in regulating reservoirs provides simple feedback of how well the supply matches the demand. Regulating reservoirs, of course, require land and are expensive. Nevertheless, they provide a simple means of achieving a high degree of flexibility if designed properly.

Today, most irrigation projects that need modernization require a blend of control strategies to match various physical, technical, and managerial constraints. For example, a main canal may be automated with some form of upstream control (perhaps with an additional element of prediction analysis). Or sometimes a main canal has upstream control on the upper 75% of the reaches, with a regulating reservoir at that point, and downstream control in the last 25% of the reaches. In short, a single control solution throughout a complete irrigation project is rarely the best design.

Certainly, computerized automation is usually more appropriate on larger canals rather than on smaller canals. The cost of a programmable logic controller, the programming, the modeling, a power supply, gate actuators, radio communications, vandalism enclosures, and so on, is about the same per structure whether the structure is large or small. Computerized automation is, therefore, often cost-prohibitive on small lateral canals.

1.6.1. Centralized versus Local versus Distributed Control

"Centralized control" in this discussion implies that the individual gate movements throughout a project are coordinated and based on water level and/or flow conditions at multiple locations. This usually requires that they are computed at a central location. "Local control" implies that each automated gate or pump acts independently from other gate or pump locations. One common usage of the term "distributed control" implies that there is local control but that the actions are monitored from a central location, and target flows or water levels can be sent to the local sites from the central location. This configuration of control and monitoring is by far the most common type of automated computerized control.

Most of the historical centralized control has been implemented on midsized and large canals, such as the California Aqueduct or the Central Arizona Project (CAP) canal. These large canals are typically much easier to operate than smaller canals because smaller canals have flow rate fluctuations of a much higher percentage, and many of the fluctuations are unanticipated. Further, the large canals often have very large terminal reservoirs, and the canals themselves may be primarily conveyance rather than distribution canals. The control also may utilize temporary canal pool storage.

In other situations, distributed control is preferred. A conscious effort is needed during the planning and conceptualization of a preferred control scheme to select the most appropriate control methods, including whether centralized or distributed control is the most appropriate.

Further guidance on which option to choose is available in many published articles, but it is beyond the scope of this publication.

1.6.2. Developing a Plan for Emergency Response and Safeguards

With electronics in the field, it is never a question of whether there will be malfunctions; it is only a question of when and how often problems will occur. Sensors and power supplies for PLCs fail. Lightning strikes. Vandalism occurs. Communications fail. Gates become stuck. There can be logic problems, even as serious as gates moving in the opposite direction than desired.

It is, therefore, essential that any automation system be implemented incrementally and have preapproved procedures on how to react to emergencies. In general, any PLC-based automation scheme must first have an excellent SCADA base station installed with excellent communications established between the base station and individual control sites. When the structures are automated one at a time, the performance of each structure can be monitored immediately from the office. Any problems can be noted quickly and resolved.

After the initial commissioning, the SCADA system can be set up to display alarms if there are control problems. It is highly recommended that all automated structures have redundant sensors (for water levels and gate positions) of different types (which fail in different modes). If the sensor readings deviate from each other, a technician can be sent out to determine which sensor is in error. While such a sensor problem is being fixed, the automation can proceed using the secondary sensor.

In addition to the SCADA system, one must always consider how various elements of the system will respond in extreme cases such as failure of the regional electrical supply. Whereas this might not affect the operation of gates (if they have solar power or backup generators, for example), large numbers of irrigation pumps may shut off, and the canal flows can quickly increase or decrease. One must also consider the possibility of a control malfunction. The need for emergency side spills does not decrease with canal automation; in fact, it increases.

1.6.3. Expanding on Historical Success with Local Automation

As of the writing of this publication, there are certainly hundreds, if not thousands, of successfully automated canal structures that use some form of PLC or electronics for controlling operations. However, these are typically stand-alone, single structures that operate independently. For example, the most common structure to automate is a regulator at the head of a canal, which is typically automated for flow control using proven technology. Caution must be exercised when consideration is given to replicating that successful technology at structures in other locations.

When a modernization project moves into the realm of more than three gate structures in series with rapid flow rate changes, the old computerized automation methods (which were simply a well-thought-out series of control actions that "made sense") cannot provide the rapid and stable control that is necessary. In projects involving the automation of multiple control structures, a completely different approach to canal automation is needed. This approach must recognize the unique hydraulic characteristics of each pool and cross regulator, plus the interactions between adjacent pools and gates. Because most people are familiar with the success of simple control logic applied to one or two gates, it is often difficult to convince them that the calibration of gate control constants with unsteady flow modeling, plus modern control logic, are needed. This publication attempts to clarify ingredients of modern canal control so that informed and appropriate decisions can be made.

1.7. WHEN TO SAY NO TO A PLC-BASED AUTOMATION PROJECT

For a modernization project of the kind addressed in this publication to be a success and to provide the benefits anticipated, all involved parties need to be dedicated and committed. This starts with district boards of directors and management and extends down through the district staff and on to the consultants, integrators, and contractors. Success requires a team approach.

Although there are benefits to be realized from increasing service levels to the water users in terms of making their farming operations more efficient, there are times when a PLC-based canal automation project should not be pursued or should be delayed. Any one of the following red flags can be sufficient to render a canal system automation project ineffective:

1. The irrigation project authorities are not committed to maintaining their canal infrastructure, as evidenced by a quick field tour that clearly indicates extensive deferred maintenance. This lack of maintenance is due to the lack of accountability and financial transparency, lack of a maintenance plan, and other reasons.
2. Budgets are inadequate to sustain the automation project. This is especially common in some projects in which donors fund the

automation and the government has no commitment to maintaining it. An annual maintenance budget of at least 15% of the cost of PLC-based automation components is a good starting point.

3. There is a rush to design and install a large modernization project within a few years, although the project staff does not have adequate experience with modernization concepts, much less with modern SCADA and automation equipment.
4. The project staff is not qualified to operate and maintain the equipment, and the project authorities are unwilling to spend enough money to train the staff properly in advance for many months or years.
5. There is a problem obtaining sufficient electronic diagnostic equipment and spare parts before the modernization is implemented.
6. There are unreasonable expectations of massive water conservation and easy operation.
7. There is a high turnover rate among irrigation project staff.
8. There is not a clear, advanced delineation of responsibilities among the designers, builders, operators, SCADA, integrators (see Glossary), and others.
9. The irrigation project managers want to use inexpensive PLCs and sensors, regardless of the requirements and capabilities.
10. There is not a clear understanding that there will inevitably be setbacks with implementation.
11. The project managers want to invent, develop, or construct their own sensors and actuators.
12. There are no qualified local integrator companies or organizations in the area.
13. There is no well-defined operational objective.
14. The people who ultimately operate and maintain the new system are not brought into early and continued discussions as important participants. Instead, all the planning and design meetings are limited to designers and high-level irrigation project officials and economists.
15. The designers have little or no experience with modern irrigation system design.
16. Hardware manufacturers promote a system based on available hardware. The common saying is then true: "When all you have is a hammer, every problem looks like a nail."

Successful PLC-based automation is complicated and expensive and requires excellent maintenance and equipment. It involves hard work, constant attention to detail, and an almost immediate response when problems arise. There is simply no shortcut to success. Everyone must understand this in advance and have the budget available to succeed.

One must ask, "Why even use PLC-based automation?" The answer is somewhat complex. First, if the same objectives can be accomplished with a simpler means of control, then by all means the simpler means should be used. However, as water control becomes more complex and as more flexibility is introduced into a system, it is commonly necessary to introduce some aspects of PLC-based automation into some points of control. In some cases, the number of control points can become quite large, if the system is larger and more complex.

But always, one must keep in mind that PLC-based automation, or any automation, is not the objective. Generally the objectives are to improve water delivery service, protect the environment, and improve conveyance efficiency. In some cases the primary objective of automation is to reduce labor requirements. A wide variety of automation tools are available as options to help accomplish these objectives.

1.8. WHAT IS INCLUDED IN THIS PUBLICATION

The remainder of this publication is aimed at the planning, design, construction, installation, and commissioning of canal automation systems that utilize state-of-the-art electronic sensors, communications gear, and control equipment to improve the level of service to the project beneficiaries and to satisfy the legal, environmental, or operational constraints imposed on the project. Although non-PLC control options are briefly described, this publication is not intended to address control structures with fixed configurations adequately such as long-crested weirs, hydraulic gates (such as those by Alsthom Atlantic), or electromechanical operational devices (such as the U.S. Bureau of Reclamation's Little Man controller). This publication presents the reader with considerable detail, thereby facilitating intelligent discussions regarding the planning, design, and implementation of a modern-day system to control a complex irrigation canal system.

1.9. REFERENCES

Burt, C. M., and Styles, S. W. (1999). *Modern water control and management practices in irrigation: Impact on performance*. Water Reports #19. Food and Agriculture Organization of the United Nations, Washington, DC.

Clemmens, A. J., and Burt, C. M. (1997). "Accuracy of irrigation efficiency estimates". *J. Irrig. Drain. Eng.* 123(6), 443–453. doi:10.1061/(ASCE) 0733-9437(1997)123:6(443).

Renault, D., Facon, T., and Wahaj, R. (2007). "Modernizing irrigation management—the MASSCOTE approach." Food and Agriculture

Organization of the United Nations, FAO Irrigation and Drainage Paper 63, Rome.

Wolter, H., and Burt, C. M. (1997). "Concepts for irrigation system modernization." FAO Water Report 12; RAP Publication 1997/22. *Proc., Expert Consultation on Modernization of Irrigation Schemes: Past Experiences and Future Options*. Food and Agricultural Organization of the United Nations. Bangkok, Thailand. November 26–29, 1996.

CHAPTER 2

PHYSICAL INFRASTRUCTURE

2.1. WORKING WITH AN EXISTING IRRIGATION SYSTEM

Before the implementation of a modernization or automation scheme, the canals and pipelines often already exist, and the focus of modernization is on how to improve control, as opposed to designing and constructing completely new canals or pipelines. Often a choice must be made between automating existing structures or constructing new structures in the canal and off the canal system. Both approaches have been successful, although there are advantages and disadvantages in either case. It is rare that completely new projects or all new canal structures are constructed. Therefore, the physical infrastructure changes with modernization are normally limited to

1. Placing motors and actuators on existing manually operated cross regulator (check structure) or flow control gates;
2. Replacing single-speed pump motors with variable frequency drive motors;
3. Improving the lining of canals;
4. Replacing existing, old check structure gates with newer structurally sound gates of the same basic design;
5. Replacing existing, old check structure gates with gates of completely different designs;
6. Replacing stop logs (flashboards or check boards) with gates, using as close to the same gate openings as possible;
7. Installing new turnout (offtake) flow control and measurement devices; and
8. Installing improved flow control and flow measurement devices in the canals.

Control structures often are called "cross regulators" or "check structures" when they are used in-line in canal systems. These cross regulators, using one of many available logics, are usually used to maintain some type of target water level in the canal either upstream or downstream of the cross regulator. Other control structures, which may look identical to cross regulators, are necessary to control discharge into canal heads or control discharge at distribution nodes.

In some countries, the physical infrastructure was standardized with little variation among all the projects within the country. Historically (before 1990), there was little computerized automation. Old designs often were based on one of two approaches:

1. *U.S. Bureau of Reclamation design manuals*. These emphasized radial gates in the main canal, with some type of manual orifice flow control for turnouts. The turnout design would be of a constant head orifice (CHO) design that combines flow control with flow measurement, or an orifice flow control with a flume or weir downstream for flow measurement. Note that the term "flow control" does not imply any automatic function; it designates a device that is adjusted to achieve the desired flow rate. The term "constant head" is somewhat confusing because if the canal water level changes, the head difference (and therefore the flow) changes across the turnout. Examples are found in much of South America, North America, and Thailand.
2. *French hydraulic designs*. These designs incorporate hydraulic gates such as AMIL, AVIS, AVIO, or MIXTE, or long-crested weirs for water level control in the canals. Turnout designs were often of the "distributor module" design. Another common characteristic with some of these designs was the use of canalettes that were half-pipe sections suspended above the ground surface on yokes. Examples are common in northern Africa and southern Europe.

2.2. CONVEYANCE SYSTEM CONSIDERATIONS

There is no single best form of canal automation. The type of canal automation that is physically possible and practical is dependent on the physical characteristics of the existing canals. The following are some examples.

1. Some systems consist of a single main canal with a single source, with branches as the water proceeds downstream. Other systems have complex water supplies such as
   a. *Conjunctive use systems*. Conjunctive use projects utilize both surface and subsurface supplies (wells). The wells may be operated for great flexibility on an hour-to-hour basis (rare) or to

supplement water during the peak water use season (typical). The wells may discharge directly into project authority canals, or they may be used downstream of canal authority control (i.e., provide water directly to users).

b. *Recirculation systems.* Many projects recirculate drainage water from open ditches that originates from lateral subsurface inflow, surface runoff from fields, and operation spills from canals. There are always questions regarding reintroducing this poor-quality water in relation to both salinity and sediment issues and how to physically incorporate these variable drainage flows into irrigation conveyance or distribution systems that are ideally responsive to downstream demands.

c. *Multiple surface sources.* Projects often have controlled and uncontrolled surface inflows at multiple locations. A single control strategy that would work well with a simple, branching canal system would be inadequate for such a system.

2. Some systems terminate at the ocean, or another salt sink, with no downstream irrigation users. Other systems have numerous irrigation projects downstream. If there are no in-stream aquatic issues immediately downstream of the project diversion point, and if the return flows from the project are high quality, it may be possible to offer very flexible water delivery service to project users by simply diverting more water than needed. Excess flows may return to the river for use downstream. In the United States, due to in-stream flow requirements, water rights, and return flow issues, this practice is much less acceptable today than 30 years ago.
3. Steep canals have fewer control options than do very flat canals. Specifically, techniques of downstream control are more difficult to apply to steep canals. The determination of what is too steep requires modeling and hydraulic characterization of the canals.
4. Water sediment load may influence allowable minimum velocities. At low velocities, sediment will drop out of the water and plug canals or pipelines. A design that provides very flexible water deliveries with constant water levels in canals can result in significant silt deposition problems. Some projects have created large, slow velocity canal reaches in which sand and silt are deposited and then removed. Other projects have large desilting basins at the inlets to canal systems. An example is the entrance to the All-American Canal in Imperial Irrigation District in California. Districts may opt to provide high flexibility along with the associated canal sedimentation problems by deliberately implementing good annual desilting programs with appropriate equipment.
5. Availability of storage is a key issue in modernization. Most discussions of canal automation assume that the water supply is flexible. That is, the flow rates into the main canals can be modified on

demand. This may be true if there is a reservoir supply or if the canal diversion is small relative to a supply river flow rate. But in some projects, the supplying rivers have no storage, and the irrigation systems are called "run of the river." As the river flow increases, the irrigation system diversions increase proportionally. In the western United States, the water available to run-of-the-river systems is modified somewhat by state-controlled water rights adjudication decisions. The timing of canal flow rate availability may have no correspondence to the temporal pattern of crop water requirements. These run-of-the-river systems must typically be designed for equitable distribution of water, rather than being designed to provide water when requested by the users. Interestingly, equitable distribution of water often requires hardware and reliability similar to more flexible irrigation systems.

6. Flow rate capacity limitations often must be eliminated before improved hardware will be effective to provide more flexible and reliable deliveries. These flow restrictions often occur at tunnels, flumes, road crossings, and so on.

2.2.1. Types of Control Structures

Check structures (cross regulators) and other control gates can be classified according to different criteria such as

- Their possible function in a canal system (upstream or downstream level control, flow control);
- Type of flow through the structure (undershot, overshot);
- Movement mechanism (mechanically or electrically actuated structures, hydraulic gates, fixed structures); and
- Sensitivity to submergence.

Thanks to their hydraulic properties, some structures also are used for discharge measurements. For automation, often it is necessary to have discharge equations that express the link among the variables of interest (e.g., water levels, gate position, or discharge). Simple relations are presented following, although this publication does not discuss the details of structure calibrations for flow measurement capabilities. Most canal check structures in an automated system are suitable for water level control or changing flow rates but are not designed to provide good flow measurement.

The hydraulics of undershot gates (orifices) and overshot gates (weirs) are described in Chapter 5. For the purposes of this chapter, the following are the main points:

1. For an orifice, the flow rate (Q) is proportional to the square root of the head difference (H) across the orifice.

$$Q \propto \sqrt{H} \tag{2-1}$$

2. For a rectangular weir, the flow rate (Q) is proportional to the upstream head (H) over the weir crest, to the 1.5 power:

$$Q \propto H\sqrt{H} = H^{1.5} \tag{2-2}$$

3. The flow area across a weir increases as the head (H) increases, but the flow area of an orifice does not change as the head (H) increases. Therefore, the percentage of flow rate change across a weir is substantially greater than for an orifice with the same percentage change in H.
4. Due to item 3, there are several fundamental principles for control structures in manually operated systems:
   a. Orifices are better for flow rate control than weirs.
   b. Weirs are better for water level control than orifices.
   c. If a weir is very long, with a small H, a large relative flow rate change will pass over the weir with a small absolute change in H.

2.2.2. Movement of Control Structures

Some control structures are moved manually, some are in a fixed position, and others are moved automatically. For PLC-based automation (automation that uses programmable logic controllers, or "computers"), exactly when they are moved (if they do move), how far they move at a time, and the observations on which the movement decision depends all rely on the adopted control strategy. For example, in canal automation the movement may be triggered by a desire to

1. Maintain a target flow rate, measured by any of a number of devices that may be upstream or downstream of the check structure, or may be measurable by the automated device itself;
2. Maintain a constant water level upstream of the structure;
3. Maintain a constant water level at some point downstream of the structure, perhaps even many kilometers distant; or
4. Pass a flow rate change in anticipation of that change being needed somewhere else in the system.

Fundamentally, those actions are accompanied by another requirement: Check structures are needed on sloping canals to keep the water sufficiently high to make deliveries to turnouts (also known as offtakes or delivery gates), regardless of the canal flow rate. In most long, sloping canals, the normal flow depth in the canal would be shallow at low flows and deep at high flows. Check structures maintain the canal water level high in most (but not all) control strategies, regardless of the canal flow rate.

Many older canal systems were designed to operate always at high flow rates. Therefore, the water levels were kept relatively high. If a new control strategy is designed, for example, to reduce diversions into the canal at certain times of the year, there may not be enough check structures in the canal to maintain sufficiently high water levels for deliveries. This is illustrated in Fig. 2-1. Therefore, the density of check structures may need to be increased if a new control system is implemented.

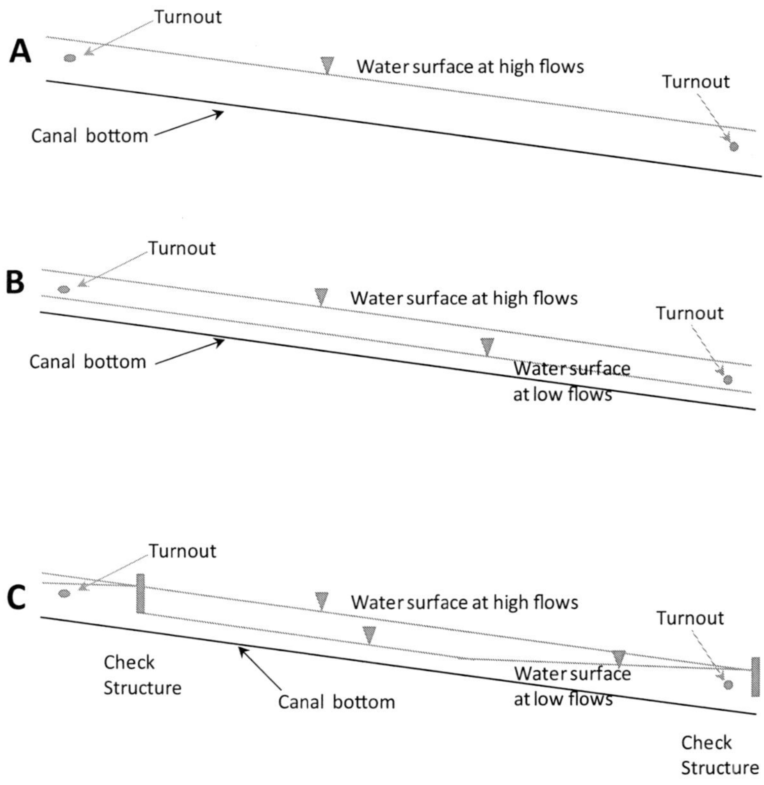

Fig. 2-1. Different water surface profiles; (A) High flow rate; (B) Both high and low flow rates with no check structures, water too low to make deliveries at low flows; (C) Both high and low flow rates with two check structures, deliveries possible at all flow rates

2.3. GATES FOR CHECK STRUCTURES

Innovation and need have provided numerous configurations of gates, some of which require electronics and others that do not. There is no single best configuration. A brief explanation of the widely available configurations is provided here.

2.3.1. PLC-Based, Electrically Moved Structures

Check structures that rely on a PLC and that are discussed in this section include

- Movable weirs (e.g., Armtec overshot, Obermeyer, Rubicon FlumeGates, or Langemann gates);
- Sluice gates;
- Radial gates; and
- Lopac gates.

It is important to note that the discussion here focuses on the hydraulic and mechanical characteristics of the various gates. This is completely separate from any automated control logic and programming that are implemented on a specific gate. If PLC-based control automation is implemented properly, there is no difference in performance between overflow (overshot) and underflow (undershot) gates, in terms of being able to maintain a desired upstream or downstream water level. A poor automation routine will have control problems, regardless of whether the flow is over, between, or under the automated gates. If complete automation failure is a concern, then flow control gates should be of an orifice configuration, and water level control gates should be of an overshot configuration.

Movable Weirs With these overshot gates, the water flows over the gate, rather than under. The height of the weir crest is adjustable. Whereas construction quality and details depend on the manufacturer or vendor, the hydraulic action of overshot has both advantages and disadvantages.

Advantages of overshot gates include the following:

1. Floating debris can pass the structure.
2. If the power fails, the weir action will maintain a more constant upstream water level than would be obtained with an undershot gate.
3. As long as there is free flow (the crest is not submerged), flow rate computations are not dependent on downstream water level

conditions. However, the flow of any gate (overshot or undershot) at a specific opening will vary depending on the downstream water level conditions if the gate is submerged.

Disadvantages of overflow gates (Figs. 2-2 through 2-5) include these:

1. Bed loads and rocks will not pass through the structure.
2. If the power fails, the weir action will not control flows into the downstream pool as well as an undershot gate of the same width would control them.
3. In a submerged flow condition, the flow rate measurement accuracy is diminished with an overshot gate.

An example is the Armtec overshot gate. The plate is hinged between two concrete sidewalls. When the gate is fully open (horizontal), the flow is uninterrupted through the structure, and the upstream water level is not affected except via the convergence of the structure itself. The weir is moved with two cables fixed on the downstream end of the plate. The cables are wound up and down with some type of electrical or manual actuator located above water level.

Rubicon FlumeGates are a recent version of the more basic Armtec overshot gates. The fundamental differences are that they have movable sidewalls, and the up–down movement is controlled positively in both directions. These are almost always automated.

Fig. 2-2. Armtec overshot gates at the Imperial Irrigation District, California, USA, in the early 1990s. The weight of the water pushes the gate down; the cables provide uplift to raise the weir. The gate is hinged at the floor

Fig. 2-3. View from downstream of an Obermeyer gate with no flow in late summer at Walker River Irrigation District, Nevada, USA. Air-inflated pillows control the angle of the rigid gate leaf opening. The gate leaf hinges at the bottom, on the floor. These gates are primarily used for coarse water level control in rivers or on dam spillways because no structure is required above them. However, monitoring and control of precise leaf positions has proven problematic in some projects

Fig. 2-4. Rubicon FlumeGate

Fig. 2-5. Langemann gates by Aquasystems 2000, upstream view. This configuration hinges on the bottom and middle. All movement of the weir crest is vertical, rather than having an arc as in the previous three gate configurations shown
Source: Photo courtesy of Mark Barnett, Western Canal Co., California, USA; reproduced with permission

Unlike the other overshot gates presented, the Langemann gate consists of two leaves hinged together. The weir is raised by vertically lifting the top of the upper leaf. The weir height is at minimum when the leaves are folded horizontally on the bottom. This design also decreases the force required to move the gates.

Sluice Gates Sluice gates (Figs. 2-6 and 2-7) typically have only undershot (orifice) action, although occasionally variations are seen with both undershot and overshot. Sluice gates are moved vertically up and down without an arc. As with bottom-hinged overshot gates, sluice gates require significant force to open and close. The force required is due to the pressure of water that pushes the movable gate leaf into the static frame, creating substantial friction when moved. Wide or tall sluice gates need to be reinforced heavily to withstand the design pressures, which can make them difficult to manufacture and install. Therefore, many automated sluice gates tend to be arranged in parallel groups of smaller—less than 1.5 m wide—gates.

Sluice gates, if placed in a structure with suppressed upstream and downstream walls and no change in the floor elevation, can be calibrated relatively easily for flow measurement. One of the newest software programs is "WinGate," being developed by the U.S. Bureau of Reclamation for this purpose. For water level control in manual mode, sluice gates are

Fig. 2-6. Automated vertical sluice gates of Left Bank Gignac Canal, France
Source: Photo courtesy of Gilles Belaud, International Center for Higher Education in Agricultural Sciences, Montpellier, France; reproduced with permission

Fig. 2-7. Automated vertical sluice gates in parallel at the Canal de Marseille, France
Source: Photo on left courtesy of Gilles Belaud, International Center for Higher Education in Agricultural Sciences, Montpellier, France; reproduced with permission

undesirable, especially if there is a large drop in water level across them. Even a small change in flow rate across the structure, if not moved, will create a relatively large change in the upstream water level.

A significant problem with sluice gates can be the amount of vertical lift required, plus the associated structures. In many countries, sluice gates

Fig. 2-8. Large motorized sluice gate in Chainat, Thailand. Equipment is difficult to access. A large structure is needed to manipulate the heavy gate

are installed in tall structures that are difficult (if not dangerous) for operators to access. These large gates also are very difficult to move without electrical actuators. Therefore, the sluice gates tend not to be moved with sufficient frequency. Figures 2-8 through 2-12 show examples of sluice gate installations.

Radial Gates Radial gates are undershot (orifice) gates that have been designed to be moved with considerably less power and effort than sluice gates. Structurally, radial gates of comparable size are lighter and easier to construct and install than sluice gates. They are common in Latin America and North America but less common in other countries. Figures 2-13 and 2-14 show typical installations.

Because almost all of the hydraulic forces of the water on the gate are borne by the rotation bearings, radial gates are easy to move. Figure 2-14 shows the importance of hooking the lifting cables to the bottom edge of the gates; the gate spools and shafts can be positioned for easy access; the tops of the gates rise well above the lifting mechanism.

Additionally, because radial gates often have little friction, they require little power to move. However, this also means that when motorized actuators are de-energized while the gate is moving downward, the gate may continue to move downward several centimeters. With automation, actuators or motors must be equipped with brakes (or special gear box designs) so that movement stops immediately when the motor contacts are de-energized. Also it is easier to provide for emergency operations of radial gates, because the lifting forces are smaller, and the actuators are designed to accommodate such possibilities.

Fig. 2-9. Manual sluice gate with large hand wheel and gear box for a small canal at Imperial Irrigation District, California, USA. The stem is covered to avoid dust accumulation. The top boards can be removed so that the sluice gate can also have an overflow weir action

Fig. 2-10. Multiple parallel sluice gates of 1.3-m width each (canal is de-watering) at the Modesto Irrigation District, California, USA. The sluice gates control the vertical opening of a hole in the bottom of the concrete structure to minimize the height of the gate

Fig. 2-11. Automated sluice gate in a small canal at Tulelake Irrigation District, California, USA

Lopac Gates Lopac gates are not widely used but are mentioned because they hinge on the sides and, therefore, pass both floating debris and bed load, as seen in Fig. 2-15. The forces on these hinges are appreciable, so this configuration generally is found in small sizes.

2.3.2. Non-PLC Controlled "Automatic" Check Structures

Although this manual of practice focuses on PLC-based control, it is worthwhile to acknowledge a wide variety of check structure designs that do not use electronic or electrical controls. Many of these gates are commonly used in new, modern installations because of their simplicity. These types of structures include long-crested weirs, hydraulic gates (e.g., AVIO, AVIS, AMIL, or MIXTE), and flap gates (e.g., ITRC flap gates, Begemann gates, or Vlugter gates). These gates, due to their simplicity, have two major disadvantages when compared with PLC-based control:

1. *Lack of flow control*: By themselves, none of these gates can deliver a desired target flow rate to a downstream canal.

Fig. 2-12. Special configuration of the shaft on a sluice gate (two gates on the left) at Merced Irrigation District, California, USA. The shaft is connected to the bottom of the gate rather than to the top. The shaft is also offset from the gate. This enables the structure to be much shorter than with a typical sluice gate. The motor and actuator are set at a level that makes them serviceable

Fig. 2-13. Three radial gates in parallel at Glenn-Colusa Irrigation District, California, USA

Fig. 2-14. Radial gates, downstream view with gate spools and shaft seen on the top of the structure at San Luis Canal Company, California, USA

Fig. 2-15. Lopac gate seen from upstream side at South San Joaquin Irrigation District, California, USA

2. *Capable of maintaining only one target depth in a canal:* The hydraulic gates described as follows only provide stable water level control at one water level. The gates themselves are so heavy that it is impractical to move them up and down in a structure.

Long-Crested Weirs Long-crested weirs have been used extensively throughout the world for simple upstream water level control. There are many details in other publications regarding crest length, avoidance of silt deposition, avoiding "floating" due to hydraulic forces, and crest height, but the weirs' lack of moving parts is appealing in some situations. The basic concept is that if the water flows over a very long crest, the depth of water over the crest is small. Therefore, a large change in flow rate will cause only a small change in water depth in the canal.

There are three most common causes of operational problems:

1. The weir length is designed without consideration of the allowable change in water level to deliver satisfactory service to upstream turnouts.
2. Flushing gates must be installed to remove silt and sand continuously.
3. The crests of the weirs are fixed, whereas it is best to have them adjustable in case of construction or design errors.

Figures 2-16 through 2-18 show example installations.

Hydraulic Gates—AVIO, AVIS, AMIL, and MIXTE These automatic gates were developed in France but have been manufactured or copied in other countries, often unsuccessfully because of lack of attention to detail. The AVIO and AVIS gates are designed for downstream control. The AVIO

Fig. 2-16. Long-crested weir that lacks silt flushing gates on downstream ends at the Solana Irrigation project in Colombia, South America

Fig. 2-17. Long-crested weir, duckbill type at Canal du Congrès, France. This design would quickly fill up with sediment if the water was not extremely clean
Source: Photo courtesy of Gilles Belaud, International Center for Higher Education in Agricultural Sciences, Montpellier, France; reproduced with permission

Fig. 2-18. Long-crested weir, ended with a movable sluice gate for sediment and algae flushing at Gignac Canal, France
Source: Photo courtesy of Gilles Belaud, International Center for Higher Education in Agricultural Sciences, Montpellier, France; reproduced with permission

is supplied by a pipeline (orifice), and the AVIS gate is supplied by an open water surface (Fig. 2-19). The AMIL gate is designed for upstream control (Fig. 2-20).

The AVIS, AVIO, and AMIL design is based on having balanced couples. The buoyant force of a float (on the downstream surface for the AVIS and

Fig. 2-19. AVIS gate at the Doukkala Project, Morocco

Fig. 2-20. AMIL gates at Government Highline Canal, Grand Junction, Colorado, USA
Source: Photo courtesy of Ram Dhan Khalsa, U.S. Bureau of Reclamation; reproduced with permission

AVIO gates; upstream on the AMIL gates) creates a couple that rotates that part of the gate upward. The weight of the gate creates another downward couple that opposes the upward couple. The gates are designed so that chambers can be filled with weights to create the exact "gate weight couple" necessary, and at the correct position in the structure. Although there is a slight control decrement (the controlled water level varies somewhat with an open versus closed gate; the larger the gate, the larger the decrement), the water level control is reasonably accurate. No electricity is needed, and maintenance requirements are minimal.

The AVIS, AVIO, and AMIL design can be applied with a relatively small head loss across the structure. Once the structures are installed, the target water level literally is fixed in concrete and cannot be adjusted. Also, the radial gate leaf must be tapered in a trapezoidal shape so that when it opens even a small amount, there is complete clearance along the sides of the gate leaf, providing no side friction. Because of this construction requirement, they require a specific concrete support structure (Fig. 2-21) and cannot be easily fit into an existing gate structure without reducing the flow cross-sectional area significantly.

Unlike for the AVIS gate, the flow through AVIO gates passes through an orifice prior to the gate leaf. Therefore, it involves a larger head loss.

MIXTE (or Vanne Mixte) gates (Fig. 2-22) were designed to provide downstream control but automatically shift to upstream control if the upstream water level drops too much because the downstream control is

Fig. 2-21. AVIS gates at a canal in Boisgelin-Craponne, France
Source: Photo courtesy of Gilles Belaud, International Center for Higher Education in Agricultural Sciences, Montpellier, France; reproduced with permission

Fig. 2-22. MIXTE ("Vanne Mixte") gate

extracting too much water. To accomplish this, there are two float mechanisms on the gates.

Flap Gates Flap gates consist of plates that can rotate around a horizontal axis. A counterweight ensures that a minimum water level is reached before the gate opens. As with the AVIS, AVIO, and AMIL design, these gates work on a principle of balanced couples. The hydraulic force on the gate causes the gate to open; the weight of the gate causes it to close. In addition, if the center of gravity is in exactly the correct location, the flap gate will maintain an upstream water level within a few centimeters.

Dutch engineers designed such gates in the 1940s for the control of upstream level in irrigation and drainage canals. Of interest are the Begemann gate (Litrico et al. 2005) (acting in free flow only), which has a simple design, and the Vlugter gate (Belaud et al. 2008), which is equipped with a float on its downstream side to limit the effects of submergence. The Irrigation Training and Research Center (ITRC) flap gate is similar to the Begemann gate. The design spreadsheet is available for download free of charge at www.itrc.org.

The primary advantage of these gates is that they are inexpensive to construct and install. Therefore, there are hundreds of these on small canals that would be prohibitively expensive to automate with upstream control using PLC-based control.

There are disadvantages of the ITRC flap gate and the Begemann gates. They require free flow. That is, the device cannot be submerged on the downstream end. The vertical height of the upstream submergence is limited to about 90 cm. At greater water depths, the counterweights become too large for practical use. Oscillations that may appear in some hydraulic conditions usually can be eliminated with steering stabilizers. Figs. 2-23 through 2-25 depict various gate types.

2.3.3. Gate Actuators

Electric gate actuators can be used to move gates up and down. There are large differences in quality, reliability, and complexity among the actuators that are available. See Figs. 2-26 and 2-27 for examples.

Fig. 2-23. Begemann gate at Hadejia Valley Irrigation Project, Nigeria
Source: Photo courtesy of Xavier Litrico, Centre R&D de Lyonnaise des Eaux à Bordeaux, France; reproduced with permission

Fig. 2-24. Begemann gate with a special "Vlugter" back to allow submergence at Hadejia Valley Irrigation Project, Nigeria
Source: Photo courtesy of Xavier Litrico, Centre R&D de Lyonnaise des Eaux à Bordeaux, France; reproduced with permission

Fig. 2-25. ITRC flap gate at San Luis Canal Company, California, USA

Fig. 2-26. Sluice gate actuators at the entrance to a regulating reservoir at Central California Irrigation District, California, USA

In locations that have excellent access to commercial actuator representatives and service equipment, commercially available packages commonly are used that provide all of the following components:

- Troubleshooting diagnostics package and display;
- Gate position sensor;

Fig. 2-27. Actuator (on right side) for a radial gate on the Government Highline Canal, Grand Junction, Colorado, USA. Note the high silt content of the water

- Motor (available with DC, or single or 3-phase AC current);
- Gear box;
- Limit switches;
- Over-torque protection; and
- Hand wheel (or nut for hand wheel connection) in the event of power failure.

In areas without ready access to excellent actuator packages, there are often local designs that perform most of the same functions. These locally designed or fabricated packages are sometimes more expensive than the commercial packages, and they do not benefit from the many years of design experience that good commercial actuator manufacturers have. Nevertheless, some of the local packages are satisfactory if the components are sized properly. Local packages rarely have any torque control. Fig. 2-28 shows an example of one such locally fabricated actuator package.

Because many gate automation projects use solar power and batteries for powering both the gate movement and PLC/RTU (remote terminal unit) units, it is important to find actuators that are efficient and that do not require a large number of amp-hours for movement. Also, the serviceability requirements and ease of DC motor adaptability must be considered. Power requirements are determined largely by the gate design.

For all gate actuators, the procedure for calibrating the position sensors must be well-defined and simple. Some automation specialists require

Fig. 2-28. Locally constructed actuator for a sluice gate at Tulelake Irrigation District, California, USA. Motor and gearbox are on the left side. Further gear reduction is accomplished with a chain and sprockets

that redundant position sensors be installed on all automated gates in case the primary position sensor fails, as well as to observe if the two sensors begin to drift apart (indicating a problem with one of them). The actuators themselves usually are not sold with redundant sensors, so the integrator must install an additional sensor.

For all installations, the following technical specifications are required:

- The torque needed for gate movement in the most difficult situation;
- Minimum resolution of gate movement, which is extremely important with modern PLC-based control that may require precise gate movements;
- Ability of the motor and actuator to stop exactly when the electrical relays are de-energized;
- Resolution for the sensors, which depends on the quality of control that is desired (for example, an 8-bit sensor generally is inadequate for most gate position sensing);
- Minimum speed of movement of the sluice gate shaft or radial gate shaft: The designer should know the maximum gate movement that the control will require during any control time step and how much time that gate movement requires; and
- Ability of each gate position sensor to retain memory of its position if there is a power outage.

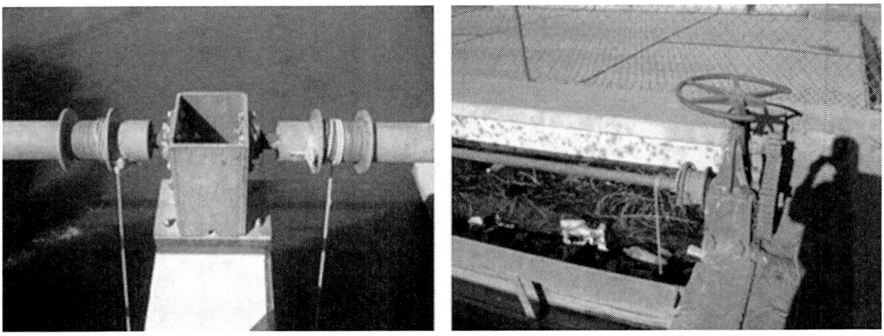

Fig. 2-29. Cable drums without grooves; the overlapping of the cable provides an inconsistent relationship between rotations of the shaft and movement of the cable

For gates that use a vertical shaft for movement, and for which the gate also moves vertically, it is relatively simple to calibrate the position sensors. For radial gates and overshot gates, however, the vertical movement of the gate is nonlinear with respect to the movement of the actuator or shaft. Therefore, the sensor reading must be converted into a vertical gate movement with a special process and formula. The relationship between the sensor raw count and the vertical position of the gate is done in the field during commissioning.

The actuator position sensor measures the movement in the actuator, not the gate itself. Therefore, for radial gates and overshot gates, it is important that the cables not overlap on themselves when the cable drum rotates. Drums should have grooves in them to provide for neat, repeatable placement of the cable during movement. Also, cable wear due to longevity or inadequate maintenance can cause errors in the position sensor reading over time. Figs. 2-29 and 2-30 show drums without and with grooves.

2.4. INSTRUMENTATION AND MEASUREMENT

The following discussion provides a cursory treatment of instrumentation considerations that are common for canal automation. Here are some important recommendations:

- Do not attempt to reinvent measurement equipment such as water level sensors and gate position sensors. This approach is almost always accompanied by eventual failure.
- Use high-quality, commercial sensors.

Fig. 2-30. Drum with grooves to ensure smooth and consistent wrapping of the cables

- Use redundant sensors of different types on all critical measurements in automation.
- Consider how the sensors will be inspected and maintained easily and how, if removed, they can be repositioned in a bracket or holder so they go back into exactly the same position as before removal.
- Keep adequate spare parts in inventory on the project, so damaged or failed devices can be replaced quickly.
- Absolutely have a well-trained, equipped, stable, motivated, and mobile staff that can troubleshoot and repair or replace problematic equipment.
- Consider the process to calibrate sensors. For example, can the gate position sensor be replaced and calibrated with water in the canal?

There are important characteristics of sensors to consider:

1. *Resolution* is the smallest change in the measured variable that can be measured. For example, if the device can measure within 0.1 mm, that is 10 times better resolution than within 1.0 mm. The resolution that can be obtained with electronic sensors is largely dependent on
   a. The number of bits related to the sensor *and* in the analog or digital converter in the PLC. Bits are the number of binary spaces allotted to a single reading. Binary numbers consist of zeros and ones only (number of possible combinations of zeros and ones = 2^{bits}). An example of a 3-bit binary number is 010. There are

$2^3 = 8$ possible 3-bit binary numbers: 000, 001, 010, 011, 100, 101, 110, and 111. A 3-bit signal would then have very poor resolution, because it could read only eight different sensor output values. If the sensor span equaled 70 cm, the analog or digital converter would have a 10-cm resolution (0, 10, 20, 30, 40, 50, 60, 70 cm). For each increase of one bit, the resolution improves twofold. Typical resolutions are 8-, 12-, 16-, and 32-bit. In general, modern automation requires at least 16-bit devices. Some of the bits are often used for "overhead" in the devices.

b. The span of sensor. This refers to the difference between a 0–30 cm sensor versus a 0–3 m sensor. For the same number of bits, the 0–30 cm sensor would have 10 times better resolution in the signal. In other words, if the expected change in water level is 30 cm, one should not purchase a sensor with a 3-m span.

2. *Accuracy* is the ratio of the error to the full-scale output or the ratio of the error to the output, as specified, expressed as a percentage. This term requires some special discussion. A device may be accurate yet provide inaccurate information in the field. For example, an accurate water level sensor may be used, but the control is typically based on water surface *relative elevation*, rather than the actual *depth of water* over a submerged pressure transducer. Therefore, field calibration is necessary to convert the accurate water depth above the transducer into an accurate relative elevation value for the water surface.
3. *Linearity* is closeness of a calibration curve to a specified straight line. Excellent sensors have excellent linearity.
4. *Hysteresis* is the maximum difference in output, at any measured object, location, or position within the specified range, when the value is approached first with increasing and then with decreasing measured object, location, or position. In other words, does the sensor give the same output at the same water level position, regardless of whether the water level is going up or down?
5. *Repeatability* is the ability of a transducer to reproduce output readings when the same actual measured location, object, or position is monitored successive times, under the same conditions, and in the same direction. This is of special concern with acoustic Doppler and transit time flow measurement devices, which have considerable "noise" in their outputs. One hundred consecutive readings over the course of 30 s may have a coefficient of variation of 1.0 or greater. Therefore, averaging techniques are required to use the output for stable control.
6. *Sensor output signal types.* Technically, a transducer has a voltage output and a transmitter has a current output. This publication refers to both transducers and transmitters as "sensors" and

sometimes the terms "transducer" and "transmitter" are used interchangeably.

a. A common potential (voltage) output is 0–5 V or 0–10 V, and the most common current output is 4–20 mA. The output is proportional to water level and usually equals 4 mA or 0 V at the lowest readable level and 20 mA or 5 V at the highest readable level. Some water level sensors can be purchased with a V, mV, or mA output. In any case, the strength of the output signal is proportional to the measured depth or water level. Although some sensors do not read the water level continuously (i.e., they have some lag time), all 4–20 mA and 0–5 V signals are continuous. When power is first supplied, some ultrasonic depth sensors and bubblers may take up to a minute to output a reading. Submersible transducers and floats output a signal almost immediately after power-up.
b. A milliamp output has proven advantages. A voltage output is more susceptible to line noise caused by motors, solenoid valves, other data lines, or other electrical devices. In addition, resistance inherent in the data cable causes a loss of voltage signal that is proportional to cable length. When the resistance in the data logger and power source are also taken into account, it is apparent that the voltage signal can diminish significantly between the sensor and the location of the signal reading. Conversely, current devices supply a constant current, regardless of the resistance. Additionally, mA cables can be run for hundreds of feet; cables for voltage signals must be very short.

7. *Communications protocols* are a set of rules and formats that determine the communications behavior of a piece of information. It allows for the meaningful exchange of information between certain electronic devices, such as a sensor and a data logger. This is the electronic language that devices use to speak to each other. The equipment on both ends of the communication must understand the same communications protocol.
8. *Temperature sensitivity*. There are three considerations regarding temperature sensitivity:
   a. All sensors have a temperature range over which the manufacturer warranties good performance. Often in irrigation projects, actual ambient temperatures (especially in poorly insulated cabinets) can exceed these limits.
   b. Some sensors tend to have signal drift as the temperature changes.
   c. Most ultrasonic sensors have built-in temperature compensation. This function corrects for the varying velocity of sound through air at different temperatures. The temperature sensor is located in the sensor. However, a sensor in an enclosure may be in a much

hotter environment compared to the cooler air near and around a water surface.

2.4.1. Water Level Measurement

Currently there are five primary technologies used for measurement of water levels in irrigation automation projects:

1. Submerged pressure transducer;
2. Downward-looking, above-the-water ultrasonic sensor;
3. Bubbler with above-water transducer (Fig. 2-31);
4. Float with cable and linear potentiometer; and
5. Downward-looking, above-the-water radar sensor (least common) (Fig. 2-32).

As mentioned earlier, the greatest distinction is often between high-quality versus low-quality sensors and the quality of installation, rather than the specific device type. There is ample commercial literature for each of the technologies. Some of the important features of the four most common technologies are noted in Table 2-1.

Location of Water Level Sensors What may seem obvious is often missed: a water level sensor should be located at the point of interest. For example, if the intent is to control a water level at a turnout upstream of a check structure, the correct location for the sensor is where the water

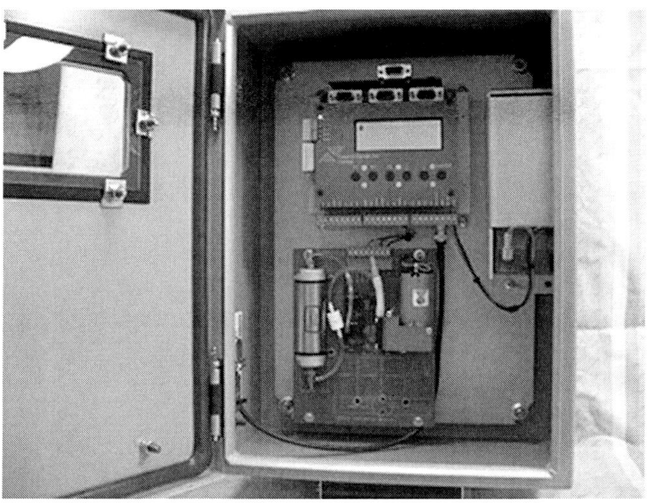

Fig. 2-31. Water level bubbler system inside an RTU panel
Source: Photo courtesy of Control Design, Inc.; reproduced with permission

Fig. 2-32. Radar level sensor at the Canal de Marseille, France
Source: Photo courtesy of Gilles Belaud, International Center for Higher Education in Agricultural Sciences, Montpellier, France; reproduced with permission

Fig. 2-33. Example of how water can converge at a gate entrance; if the sensor measures the water level at or within the entrance, the depth is quite different from that of the upstream pool. This can be a problem with pre-constructed units that have a water level sensor built into the gate structure

level is about the same as at the turnout. It is often convenient to place a sensor on the structure immediately upstream of a gate, but the water level at that point may vary tremendously depending on the velocity head of the water at that point (Fig. 2-33). Downstream, the water level sensor

Table 2-1. Typical Differences between Four Most Common Water Level Measurement Technologies

	Technology			
Consideration	Submerged pressure transducer	Downward-looking, above-the-water ultrasonic/radar sensor	Bubbler with above-water transducer	Float in a stilling well with cable and linear potentiometer
Sensitivity to silt in water	High	None	None	Medium—can change the weight of the float
Sensitivity to freezing	Can be high	None	None—transducer does not touch water	None
Requires stilling well	No	No	No	Yes
Temperature sensitivity	Small	Can be high with some models (better with radar)	Small	Small
Ease of installation and calibration	Simple	Simple	More parts than others	Ease depends on design

Fits in tight conditions	No problem	Depends on angle and sensitivity to irregularities on side of stilling well	No problem	Requires large float (min 25-cm diameter) to minimize hysteresis
Power requirement	Minimal	Minimal with some designs; high with others	Low with some designs; high with others	Minimal
Other points		Has a blanking distance; needs to be above highest water level by some distance	There are large differences regarding how often air purges the air line, and the energy and reliability	
Sensitivity to floating debris and foam	None	Very high	None	Minimal

should be located far enough away to be out of the turbulence, but not so far as to make the canal's friction losses significant.

Stilling Wells With the exception of the float with cable-type sensors, sensors usually do not require stilling wells for installation. However, stilling wells on the banks of the canals allow for easy access and some protection against vandalism or damage during canal cleaning.

Stilling wells were traditionally used for exactly what their name implies: to still or dampen wave action so that a staff gauge reading could be made accurately without having the water level bounce up and down frequently. Therefore, published stilling well construction dimensions focused on having long, small-diameter access tubes between the canal and a large-diameter vertical stilling well (Fig. 2-34). Installation of water level sensors for gate or pump automation in this type of stilling well can be disastrous. There may be too much dampening of the canal water level response, and it is possible for the water level in the stilling well to be rising at the same time that the water level in the canal is dropping.

Fig. 2-34. Very small access tube entrance to a stilling well; although the hole entrance is flush with the wall (as it should be), the diameter is too small. This tube needed to be enlarged prior to automation

Fig. 2-35. Stilling well to the side of an automated canal structure at a point of low velocity, as seen with dry canal; no access tube is used, and a staff gauge to the side allows for easy confirmation of SCADA readings

Therefore, the following are guidelines for stilling well construction:

1. Install them at the correct location (Fig. 2-35).
2. Ensure that the stilling well access tube is absolutely flush with the canal wall and not protruding into it.
3. Choose a stilling well access tube with a diameter no smaller than 25% of the diameter of the stilling well if canal levels change quickly. Stilling wells with electronic water level sensors are more access wells than stilling wells because the electronics can do the averaging.
4. Connect the stilling well access tube at least 30 cm above the bottom to allow for sediment deposition, and not at the bottom.
5. Mount the stilling well access tube higher than the bottom of the canal, because too much sand and silt would otherwise enter the stilling well.
6. Always have a plan for how the stilling well and access tube can be flushed periodically.
7. Provide ventilation on the top lid.

2.4.2. Gate Position Sensors

Much of the selection of gate position sensor type depends on the availability of commercial equipment and the ease of installation. Often with

Fig. 2-36. Encoder sensor (in box on right) attached to the end of the shaft to measure shaft rotation on radial gate. Encoders should retain memory of their position if power fails

automation, there are redundant sensors of two different types. Currently there are four primary technologies used for measurement of gate positions in irrigation automation projects:

1. Encoders that measure the rotation on shafts such as for radial gates (Fig. 2-36); such encoders should always be "absolute" encoders because absolute encoders "remember" where they were in the event of a power failure.
2. Ultrasonic sensors that are rigidly mounted on a gate frame and detect the distance between the sensor and some moving part of the gate; these are used most often on sluice gates (Fig. 2-37).
3. Linear potentiometers with a cable (Fig. 2-38); one end of the cable will be mounted to a movable part on the gate, while the other end is attached to the linear potentiometer. These are used on sluice gate shafts, on radial gate shafts, and sometimes attached to the leaf of a radial gate.
4. Inclinometers, which are used on a portion of a radial gate frame to measure the angle of the gate frame. These are fairly unusual. Although some people have had success with these, they are often dangerous to access and can provide variable results.

2.4.3. Flow Measurement in Canals

There are dozens of flow measurement techniques. All require proper construction and installation and must be installed and used in suitable

Fig. 2-37. Ultrasonic sensor (top center) pointed down to flat horizontal plate (bottom center) attached to sluice gate at Alta Irrigation District, California, USA

hydraulic conditions. For almost all devices, an ongoing maintenance program is needed to ensure accuracy and repeatability. Calibration often is necessary to establish or adjust the stage or discharge relationships. This requires punctual discharge measurements repeated for different hydraulic conditions. The discussion following is in the context of some key points related to automatic control of flow rates.

There are four commonly used methodologies for continuously measuring flow rates in irrigation channels:

- Rated sections,
- Calibrated devices and check structures,
- Flumes and weirs, and
- Acoustic Doppler velocity meters (ADVMs) and transit-time meters.

There are major differences between the four measurement techniques in regard to how quickly the flow rate will stabilize over or through the device after a flow rate change is made.

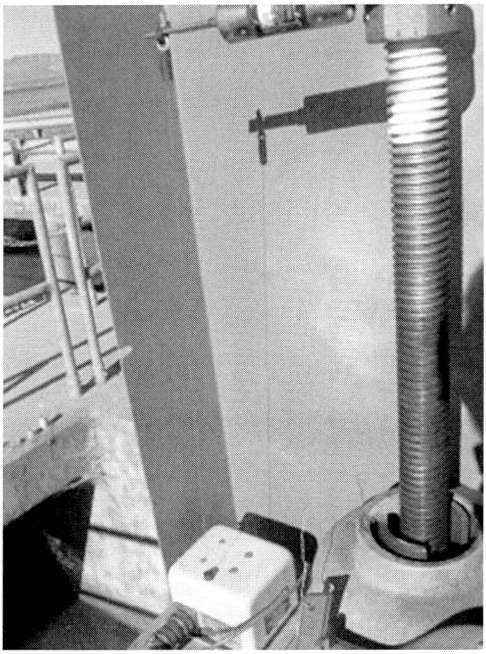

Fig. 2-38. Linear potentiometer and cable used to measure height of sluice gate shaft. These tend to have some temperature-related drift

Rated Sections Rated sections can be quite accurate if the hydraulic conditions (canal geometry and roughness) do not change over time. Accuracy also depends on not having any backwater influence from downstream structures. This qualification eliminates them from most canal applications. Also, vegetation can affect the stage–discharge relationship, even in the case of canals lined with concrete. Macrophytes are able to develop with limited amounts of sediment deposits, whereas algae and bryophytes, without roots, are able to colonize the banks (Fig. 2-39).

Assuming that the hydraulic conditions are correct, rated sections still pose a problem for rapid automation (flow stabilization in less than 30 min) because there is often a significant lag time between a change in the flow rate at the upstream check structure or dam and the water level stabilization at the rated section. This is because a long section of canal must slowly fill up or empty out between the two steady-state conditions.

Calibrated Gates Any gate can be calibrated to determine the relationship between flow rate and upstream water level, opening, and

Fig. 2-39. Bryophytes on canal bank and bottom at Gignac Canal, France. Variations of Manning's roughness coefficients reach 30% between spring and summer
Source: Photo courtesy of Gilles Belaud, International Center for Higher Education in Agricultural Sciences, Montpellier, France; reproduced with permission

downstream water level (see Eqs. [2-1] and [2-2]). Even so, the following are problems that are typically encountered:

1. Real irrigation canal systems are not research laboratories. Calibration is necessary, but typically operators do not allow people to change canal flows arbitrarily just to calibrate the flow rate. Therefore, it may take years to develop field calibration curves.
2. It can be time consuming to verify flow rates with current meters.
3. Gates often are installed in parallel, and the entrance conditions are different if one gate is open versus multiple gates.
4. With high submergence conditions and very open gates, calibration is often only marginally accurate.
5. During flow calibration it often is difficult to keep the flow rates steady.

Some of these problems can be minimized in a new project by observing these rules:

1. If all gates are identical in all aspects—including approach conditions, floor raises or drop, inlet protrusions, and discharge

conditions—then only a few of these gates need to be calibrated. That calibration then can be used on other identical structures and identical installation conditions. This requires that the entrances of parallel gates be designed so that the hydraulic conditions at individual gate entrances do not change regardless of the number of gates that are open at once.
2. The gates must have the same inlet and outlet conditions, regardless of whether one gate in a structure is being operated or all the gates are being operated together. This requires the installation of fairly long inlet walls that separate each individual gate.
3. The gates should be constructed so that they always operate in either a free flow or submerged condition, but not so that they might switch from one hydraulic condition to another.

Any gate that operates in a submerged condition requires multiple position adjustments over time to maintain a desired target flow rate, even after the target flow rate is first achieved and the upstream level remains constant. This is due to the rising or lowering of the downstream water level over time, as the hydraulic conditions change in a channel.

Flumes and Weirs Flumes and weirs should be designed and constructed according to standard guidelines. The publicly available program WinFlume, if used properly, is an excellent tool for broad-crested weir design (www.usbr.gov/pmts/hydraulics_lab/winflume/). The idea of using a broad-crested weir, another flume design, or another weir (such as sharp-crested) is that it will be a standard structure, and, therefore, the standard published stage or discharge relationships can be used. There should be no need for calibration of the device. Following are some typical problems that eliminate this theoretical advantage:

1. The designer did not consider downstream conditions, and the structure is submerged during part of the year. This may occur at either high or low flows, depending on conditions.
2. Attention was not paid to the approach velocity. Flumes are sometimes built with excessively high upstream Froude numbers, and weir discharge equations typically do not account for entrance velocities.
3. The as-built dimensions are not measured and compared to design specifications.
4. Sharp-crested suppressed weirs are not properly aerated.
5. Maintenance is lacking. Two common problems are sedimentation upstream of the device, or growth of algae on the flume (Fig. 2-40). Algae growth can be minimized by the application of special paints

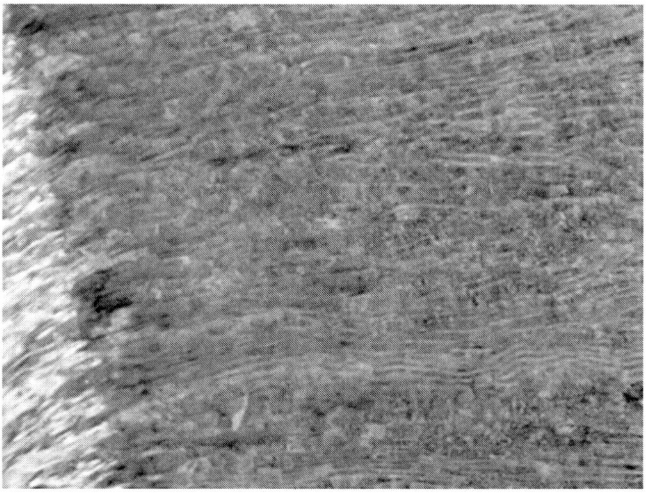

Fig. 2-40. Algae growth 3 cm deep on the crest of a broad-crested weir

Fig. 2-41. Painted broad-crested flume to minimize algae growth; flow rate capacity is 3 CMS (100 CFS) at Imperial Irrigation District, California, USA

on a very smooth flume surface. Extreme sedimentation can sometimes be dealt with by using a special flume design that does not rise above the floor of the channel. Figures 2-41 and 2-42 show techniques to minimize these problems.

Fig. 2-42. Flume with minimal rise in the floor to pass sediment

There are, however, major advantages of flumes and weirs:

1. If designed, installed, and maintained properly, there is no need for calibration with current meters; many flumes can be calibrated with as-built dimensions using a computer program.
2. Flumes and weirs located downstream of and close to a flow control gate provide almost immediate stabilization of the new flow. This is because these are critical flow devices, and the filling up action of the pool downstream of the flume or weir never affects the water level upstream of a properly designed weir or flume.

Acoustic Doppler Velocity Meters (ADVMs) and Transit-Time Meters These devices often are used in large canal sections to avoid the construction costs of installing large flumes or weirs, or in situations with very dirty water or no available head loss (which is required on flumes and weirs). There is ample commercial and technical literature available for these devices. When using these to measure the flow rate for automatic flow control, the following considerations must be given:

1. All of these devices only sample the velocities in a small percentage of the channel. Therefore, they generally must be calibrated. This, in turn, means that the hydraulic conditions around the device must stay the same all the time, and the device must be replaced in exactly the same place as before removal for maintenance (Fig. 2-43). All of this costs money; often, a concrete-lined canal section must be

Fig. 2-43. Secure frame that allows the ADVM to be removed from the channel for maintenance and then replaced at the original depth and aspect

Fig. 2-44. Special section of concrete lining into which an ADVM is placed; the bridge is used for current measurement with a current meter to calibrate the ADVM

installed in the area to keep the channel in constant shape and free of weeds (Fig. 2-44).

2. Calibration means that at least 10 to 20 widely spaced values of flow rate and sensor readings must be gathered. This is very expensive

Fig. 2-45. Transit time measurement device; beams are transmitted across the channel

and time consuming, and operators may be reluctant to adjust canal flow for calibration runs.

3. There are both horizontal beam and vertical beam units for ADVMs. It is not recommended that horizontal beam units be used if the water depth varies appreciably. With transit time devices, multiple beams must be used if the depth changes (Fig. 2-45).
4. Bottom-mounted (vertical beam) units for ADVMs can be installed in specially designed cross sections that ensure that velocity profile of each vertical slice of flow is similar to other vertical slices (Fig. 2-46). If this is the case, the "sample" velocity measurements are indeed the "average" velocity measurement, and calibration can be eliminated (Howes et al. 2010).

One characteristic of many acoustic devices is that there is a lot of "noise" in the flow rate measurements, most likely because of the microturbulence that exists in a flow area. As a result, flow rate adjustments must sometimes be based on 15- to 30-min averages, as opposed to control with nearby flumes or weirs, with which flow adjustments can be evaluated within 1 to 2 min.

2.5. PUMPS

Pumps are widely used in irrigation schemes throughout the world. Sometimes they discharge into distribution pipelines, and in many other

Fig. 2-46. Special entrance walls to ensure uniform velocity profiles at San Luis Canal Company, California, USA
Source: Photo courtesy of Alejandro Paolini; reproduced with permission

cases they are strictly for conveyance such as pumping from a river to a canal or from a canal into a reservoir. They are an integral part of many automation schemes that must supply or accept water at variable rates, which can change from minute to minute without advance notice.

There are many books on pumps and pump controls, but certain aspects are particularly important for many irrigation projects. What is unusual for irrigation applications that discharge into canals is that the control of the pump discharge is often quite complex; the sophisticated control logic that is discussed in this manual under the context of canal gates is identical for pumps. The only difference is that it is the pump discharge that is varied, rather than the opening of a canal gate. This type of control is entirely different from typical civil engineering projects in which pumps discharge into pressurized pipelines, or just work on on/off switches (similar to the Little Man controllers, discussed in a later chapter).

Variable frequency drive (VFD) controllers on pressurized pipelines typically do not need hydraulic modeling to determine the control constants. The resident proportional integral derivative (PID) control functions that are standard in many VFD motor control panels are simple but extremely easy to tune with just a few iterations of trial and error.

For most canal level control, though, the pump flow rate control algorithms are much too complex to be handled by the simple resident PID controller. A separate PLC/RTU is needed to process the variable speed

Fig. 2-47. Recirculation spill from a pool downstream of lift pumps, back to the inlet pool. This maintains the water level in the downstream pool, even though the pump flow rate is excessive

information and water sensor data, determine pump speeds and pump sequences, and transmit it to the VFD controller.

2.5.1. Variable Flow Control

There are four techniques that are typically used in irrigation district pumping plants to fine-tune flow rates:

1. *Bypass excess flow rate* (Fig. 2-47). The pump is run at a constant flow rate, and any extra water is recirculated. The recirculation flow can be controlled with a modulating valve on the pump discharge, or it can be accomplished in a spill at the head of the canal. This technique can be very simple with a spill. It results in a waste of energy, but there is no choice with a single-speed axial or mixed-flow pump. Those pump designs, used for relatively low lifts (up to 10 m or so), cause the motor horsepower to rise significantly if a pump discharge valve is closed down (modulated) to reduce the pump discharge flow rate.
2. *Modulate a valve on the discharge side of the pump* (Fig. 2-48). As mentioned, this is not useful for mixed-flow and axial-flow pump designs, but modulating a valve on a radial-flow pump requires less input horsepower than recirculating the flow. The valve can be equipped with an actuator that is automated. The only concern with

Fig. 2-48. Modulated valves (followed by check valves) on the discharge of each pump at Orland-Artois Water District, California, USA

vertical pumps is that single-speed vertical pumps that are modulated may have failure problems with the thrust bearings on the motors, due to the high thrust that the impellers develop. Appropriate thrust bearings should be installed; this should not cause problems if the thrust condition is calculated in advance of ordering the pump.

3. *Sequence multiple single-speed pumps* (Figs. 2-49 and 2-50). A large number of parallel pumps, sequenced in series, gives a relatively fine adjustment of flow rates. Sequencing is done when the canal water level has some allowable fluctuation. Conceptually, the operation is simple: a higher-than-target water level causes a pump to turn on, and a lower-than-target water level causes a pump to be shut off.

However, there are appreciable details to consider. For example, there must be sufficient variable storage in the canal pool between the high and low levels for a single pump to avoid more than about five pump starts (of an individual pump) per hour (the exact number of starts depends on the size of the motor). This helps avoid overheating of the motor. The appropriate equation is

$$V = 900\,Q/n \tag{2-3}$$

where V is the cu ft or cu m of pool storage (variation) needed, Q is the pump flow rate (CFS or CMS), and n is the number of starts per hour.

Fig. 2-49. Multiple pumps in parallel at Berrenda Mesa Water District, California, USA

Fig. 2-50. Pumping station at Torres de Segre, Spain
Source: Photo courtesy of Gilles Belaud, International Center for Higher Education in Agricultural Sciences, Montpellier, France; reproduced with permission

4. *Use variable speed pump drivers.* There are two typical configurations:
   a. All the pumps are equipped with variable speed drivers; drivers are engines or variable frequency drive controllers for a special electrical motor. This configuration is the easiest to automate. It

Fig. 2-51. Three identical pumps controlling the discharge from a regulating reservoir at Ingomar Reservoir, Central California Irrigation District, California, USA. The motor on one of the pumps is oversized; it is operated at a slightly higher Hz than the others at maximum speed and is controlled by a VFD controller (VFD not shown)

is also highly recommended for any situation in which the pumps start up against a pressurized pipeline in order to avoid water hammer problems.

b. One or two of the pumps have a variable speed, and the others have fixed speeds (Fig. 2-51). This is applicable in the case where each pump has its own independent discharge pipeline into the canal.

2.5.2. Tips for VFD Applications

Following are a few of the rules for variable frequency drive (VFD) applications:

1. Only use high-quality controllers. If the specification details are unclear, find someone who knows them. Never buy based on price alone. There are huge differences between controllers in terms of
   - Quality of the power that they send to the motors;
   - Length of cable that can be run between the motor and the controller;
   - Efficiency, which is very important in terms of the need for a panel cooling system;
   - Temperature rating;

- Electrical harmonic generation;
- Audible noise; and
- Sensitivity to variations in the supply voltage (a major consideration for many remote irrigation projects).

2. If there is only one VFD in a parallel pumping plant, the VFD controller must be assigned to the pump with the highest flow rate.
3. The pump performance must have a continuous downward slope (no dips) from zero flow to maximum flow. If the manufacturer cannot supply a reliable pump curve down to a flow rate of zero, do not use that pump. This is critical for axial-flow and mixed-flow pumps to avoid unstable control.
4. "Inverter duty" motors must be used. These usually have the following differences when compared to regular motors:
   - Windings are created using vacuum impregnation that is intended to eliminate air pockets in the systems. Air pockets can inflate as the temperature increases.
   - Magnet wire is coated with high dielectric strength film.
   - Wire size is increased.
   - Some manufacturers add phase and grounding protection using stronger dielectric materials.
   - Insulating material has a higher temperature rating.
   - Rotor bars are modified to increase the surface area and decrease slip.
   - Motors greater than 75 kW require the addition of an insulated bearing carrier installed on the upper bearing and a shaft grounding ring above the bottom guide bearing.
   - Motors of NEMA "Design A" and "Design B" designations are required for inverter-duty applications. One must specify that the motors meet or exceed the NEMA MG-1 Part 31 Standard to obtain a premium efficiency inverter duty motor.
5. The efficiency of a motor operating under variable loads and variable frequencies is the same as a similar single-speed motor operating under the same load.
6. The kW usage for lift pumps with variable frequency drives is not even closely approximated using the affinity laws if there is any elevation involved in determining the total dynamic head of the pump. In other words,

$$\frac{\text{kW1}}{\text{kW2}} \neq \left(\frac{\text{RPM1}}{\text{RPM2}}\right)^3 \tag{2-4}$$

Rather, the kW usage must be determined by examining how the pump curve crosses the system curve at various revolutions per minute (RPMs).

7. Proper cooling of the panel is extremely important and often can be accomplished only with an air-conditioning unit or newer heat sink designs that draw the heat from the panels.
8. As a rule of thumb, the VFD controller must be sized one nominal size larger than the motor that it controls.

2.6. REGULATING (BUFFER) RESERVOIRS WITH THE IRRIGATION SYSTEM

Regulating reservoirs (also known as "balancing" or "buffer" reservoirs) are commonly used in western United States irrigation district modernization projects with canal systems (Fig. 2-52). Some districts have about a dozen or more reservoirs already installed.

2.6.1. Feast or Famine Effects

Most irrigation canal systems are designed with upstream control. Under this method of control, the flow rate into the head of a canal system is set to the anticipated demand downstream. If the flow rate into the head of the canal is too high, there is operational spill at the tail end (most downstream end) of the canal. If the flow rate into the canal is lower than the demands, there is a deficit at the tail end. This phenomenon is often called "feast or famine" at the tail end.

The causes of feast or famine are many and include the following:

1. As more drip or micro systems are used, and as other field irrigation practices are improved, there is a demand by farmers for more flexibility of water delivery. This flexibility makes it more difficult for canal operators to match the canal flow rates precisely to the demand.

Fig. 2-52. Example reservoirs: Waldron Regulating Reservoir at Fresno Irrigation District (left), California, USA, and Main Reservoir (canal on right) at Henry Miller Reclamation District (right), California, USA

2. A flow rate change made almost instantaneously at the head of a canal arrives at a point several miles downstream after a considerable amount of time. Moreover, the flow rate change arrives gradually. It may take an hour (or several hours in a large canal) for a complete flow change to stabilize at a downstream point in a canal.
3. Flow rate measurement is imprecise. In the field, ±6% accuracy is considered very good.
4. There are numerous reasons that turnout (delivery gate) flows change unexpectedly, including fluctuations in the canal water level or fluctuations in the pressure or water level downstream of the turnout.

Without regulating reservoirs, feast or famine typically is handled in one of three ways:

1. It just exists. Farmers at the tail ends of canals receive the worst service.
2. The irrigation district makes a point always to deliver more water than is needed. Therefore, spill occurs all the time. This is a convenient way to provide excellent flexibility to users in a very simple manner.
3. Spill is minimized by using canal pool storage for the excesses or deficits. This means that the canal water levels constantly fluctuate and provide a fluctuating flow rate to the turnouts. These fluctuations also contribute to canal breaks and rodent damage to canal banks. In addition, oversized canals are more costly, and sedimentation issues can be aggravated.

2.6.2. Advantages of Regulating Reservoirs

Regulating reservoirs, with a classic location of two-thirds of the distance down a canal, offer the following advantages:

1. They greatly reduce canal spillage.
2. The canal operation becomes much simpler. The water master determines the change in water orders and looks at the water level in the reservoir. If the reservoir is too high, a flow rate change that is slightly less than the change in water orders is made at the head of the canal. There is no need constantly to adjust inlet flow rates to match flow rates; once or twice a day at the canal head is often sufficient.
3. Reservoirs allow for flexibility of water delivery upstream, while the canal pool water levels upstream can remain fairly constant. All of the fluctuations in flow rate show up at the regulating reservoir.

4. Reservoirs allow for more flexibility of water delivery downstream, in one of two ways:
   a. If the canal downstream of the regulating reservoir is operated on downstream control, the flow rate entering downstream of the reservoir will match the actual user demands precisely.
   b. If the canal downstream of the regulating reservoir is operated using upstream control, the downstream canal sections, for purposes of control, are only about one-third as long as without a reservoir. This allows operators to respond quickly to flow changes downstream. Also, these last sections of canal no longer absorb the fluctuations that originated in the upper two-thirds of the canal.

2.6.3. Regulating Reservoir Volumes

The volume of a regulating reservoir depends on three primary factors:

1. *Availability of land.* Although the formulas that follow provide criteria for sizing a reservoir, in the end, an irrigation district must find a willing seller of land in about the correct location.
2. *Cost of land and construction.* A regulating reservoir, plus its inlet and outlet structures, may easily cost $US1–3 million.
3. *Hydraulic requirements.* In theory, the required reservoir volume is calculated based on assumptions, including
   a. Discrepancy of actual flow rate in the canal versus the total flow rate being delivered. This discrepancy pertains to the flow rates in the canal sections both upstream and downstream of the reservoir. For purposes of illustration, assume that the discrepancy is 5% and the total flow is 35 CMS.
   b. Time required for a compensating flow rate change at the head of the canal to travel to the reservoir. For illustration, assume that this travel time is 6 h.
   c. Time required for operators to notice the flow rate discrepancy and react. Because discrepancies often occur during the evening, this reaction time may easily be 12 h.
   d. Ideally the reservoir would be half full before any flow discrepancy begins. It could either supply a deficit or excess discrepancy. The plus/minus factor is, therefore, 2.0.

$$\begin{aligned}\text{Volume needed} &= \text{Plus/minus factor} \times \%\ \text{discrepancy}/100 \\ &\quad \times \text{canal flow rate} \times (\text{travel time} + \text{reaction time}) \\ &= 2.0 \times 0.05 \times 35\ \text{CMS} \times (6+12)\ \text{h} \approx 0.2\ \text{MCM}\end{aligned}$$

Typical regulating reservoir volumes in California, for instance, vary from about 50,000 cu m to 320,000 cu m. Before the first reservoir is constructed in a district, the district personnel tend to think that the initial volume estimate is high. After a couple of years, the operators want more and larger reservoirs.

2.6.4. Sizing the Flow Rates for Reservoir Inlets and Outlets

The inlets to reservoirs typically are sized for greater flows than are the outlets because

1. The inlet to a reservoir is typically gravity flow, and the outlet often requires a pump. It is relatively inexpensive to "oversize" a gravity inflow structure as compared to a larger pumping facility.
2. Sudden rainstorms cause farmers to shut off their irrigation deliveries, which creates the requirement of high flow rate capacities into the reservoir.
3. In a modern irrigation system, it is very important to provide farmers with flexibility to shut off their water without advance notice. This allows them to conserve water. It is less important to provide the flexibility to open turnouts at any time without advance notice.

If it is anticipated that there might be a 20% discrepancy in volume during the day, this might translate to a 40% instantaneous change in flow rate.

2.6.5. Reservoir Control Considerations

There are numerous configurations of control schemes into and out of reservoirs. The exact scheme used typically depends on factors such as

1. *Topography*. For example, in some situations, it is possible to use gravity flow both into and out of a reservoir.
2. *Depth and height of a reservoir*. Some reservoirs can have gravity flow in at times, but the reservoir may also depend on pumping into the reservoir at other times.
3. *Simplicity*. Programmable logic controllers (PLCs) for pump and gate operation are standard in some projects but are too complex for others.

The control sketches on the next few pages assume that the canal control restarts downstream from the reservoir with flow control at the reservoir, that is, the canal will continue to operate on upstream control upstream from the reservoir. In some cases, the canal restarts with downstream water level control rather than flow control. The following symbols are used in Figs. 2-53 through 2-63:

▷	Upstream Water Level Control Device
▭	Flow Measurement Device
▯	Flow Control Device
○	Pump
T	Target Flow Rate
$\tilde{Q}$	Variable Flow Rate
$\bar{Q}$	Constant Flow Rate

When talking about control considerations, the following are generally true:

1. It is desirable to have the excess only from a portion of the canal flow enter a reservoir to minimize the amount of sediment that is deposited in the reservoir.
2. If there is no significant drop available in the canal, the reservoir control is usually designed to maintain a fairly constant water level in the adjacent canal pool (Fig. 2-53). A pump is needed in at least one direction of flow into or out of the reservoir.
3. If there is a significant drop available in the canal, it may be possible to have gravity flow both into and out of the reservoir (Fig. 2-54). This configuration typically has two flow control points; most of the flow rate is controlled at a check structure in the canal. The flow into the reservoir is variable, the flow out of the reservoir is relatively

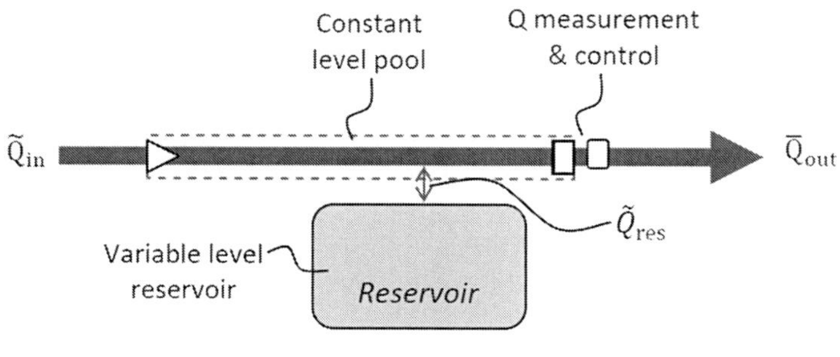

$$\tilde{Q}_{res} = \tilde{Q}_{in} - \bar{Q}_{out}$$

Fig. 2-53. Reservoir control with no significant drop in the canal

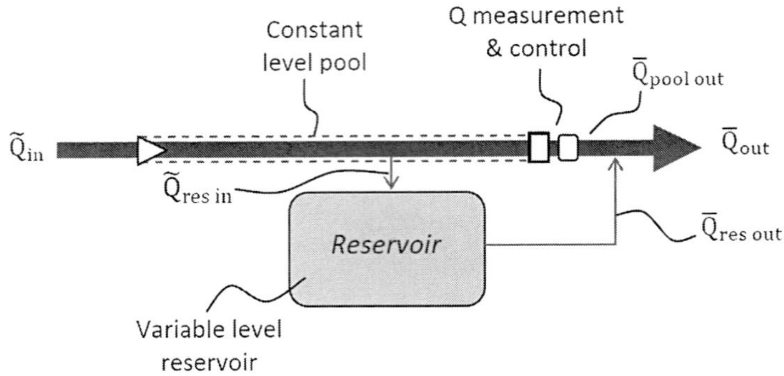

$$\tilde{Q}_{res} = \tilde{Q}_{in} - \bar{Q}_{out}$$

Fig. 2-54. Reservoir control with a significant drop in the canal

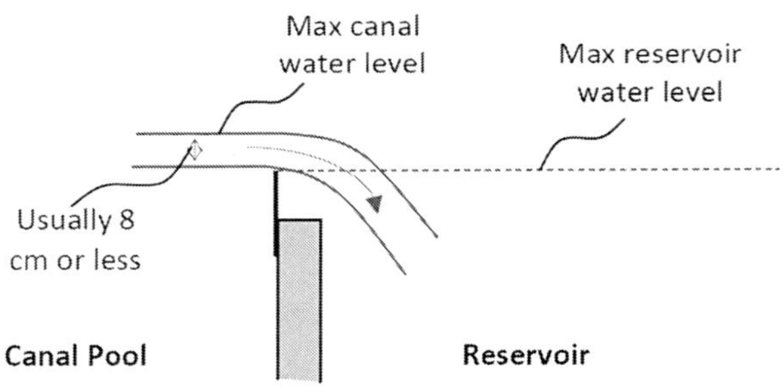

Fig. 2-55. Spill into reservoir via a long-crested weir

small, and the water level in the reservoir varies. The gravity flow into the reservoir provides automatic upstream water level control in the adjacent pool.

4. To conserve pumping energy requirements, generally it is more beneficial to use gravity *into* a reservoir and pump *out*.

Passing Excess Canal Flows into a Reservoir There are various means of providing upstream control of the canal pool that is adjacent to the regulating reservoir when excess flows arrive. These include

1. Spill into the reservoir via a long-crested weir (Fig. 2-55). If the weir is long enough, the rise in water level in the reservoir can be maintained at 8 cm or less. The reservoir can fill to the crest of the long-crested weir.

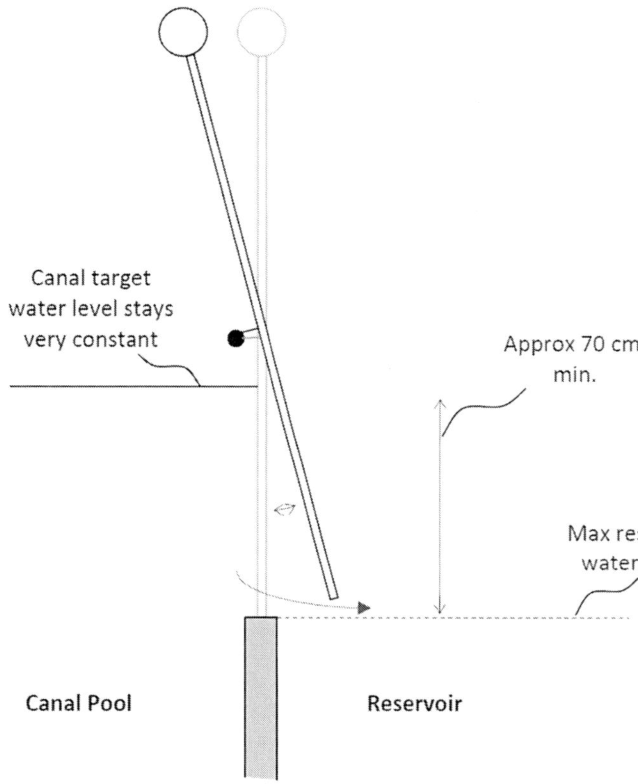

Fig. 2-56. Spill into reservoir via an ITRC flap gate or Begemann gate. This would rarely be used

2. Spill into reservoir via an ITRC flap gate (Fig. 2-56). This technique maintains a tighter control of the canal pool water level but about 70 cm of storage is lost in the regulating reservoir. That is very expensive storage, so this is not usually an acceptable option.
3. Gravity flow into the reservoir using a PLC-controlled gate that provides upstream water level control on the adjacent canal pool (Fig. 2-57). If the control is good, it provides the best combination of good canal pool level control and minimal loss of regulating reservoir storage. It is also much more complicated than a long-crested weir entrance.
4. Pump into the reservoir via a pump. This is necessary if the reservoir water level is higher than that of the canal pool. The control can be complex or very simple. The simplest controls (no VFD, no transducers) provide the most canal pool fluctuation. Several variations are listed in Table 2-2 and described here:

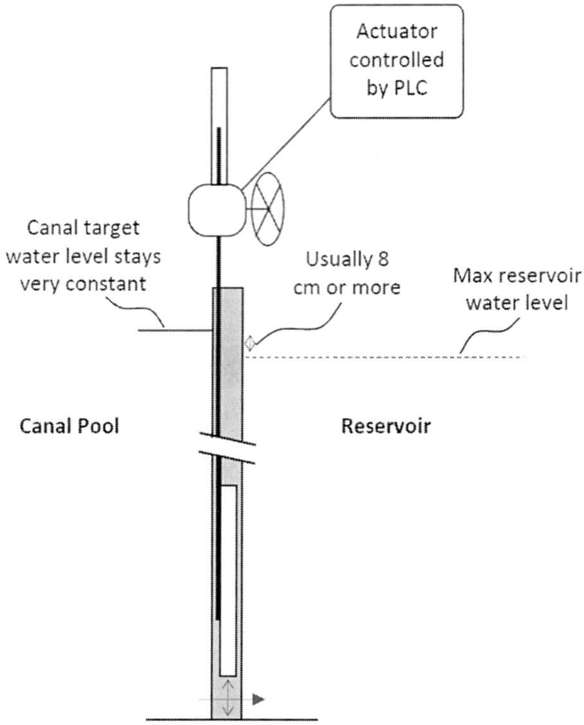

Fig. 2-57. Gravity flow into reservoir using PLC-controlled gate

Table 2-2. Simple Summary of Pump Control Options

Pump option	Tightness of water level	Complexity
A	Best	Most
B	Middle	Middle
C	Worst	Least

Option A: Electronic level sensor, PLC, VFD, and complex logic (e.g., PI). Pumps are sequenced, one at a time, to maintain target within a close tolerance (3 cm). Fig. 2-58 shows the configuration.

Option B: Electronic level sensor, PLC, and single-speed pumps (Fig. 2-59). When water level rises to "on," one pump turns on. If the flow is sufficient, the water level will drop, and the pump shuts off at target. If insufficient, the water level continues to rise above "on" and after 1 min, turns on a second pump, and so on.

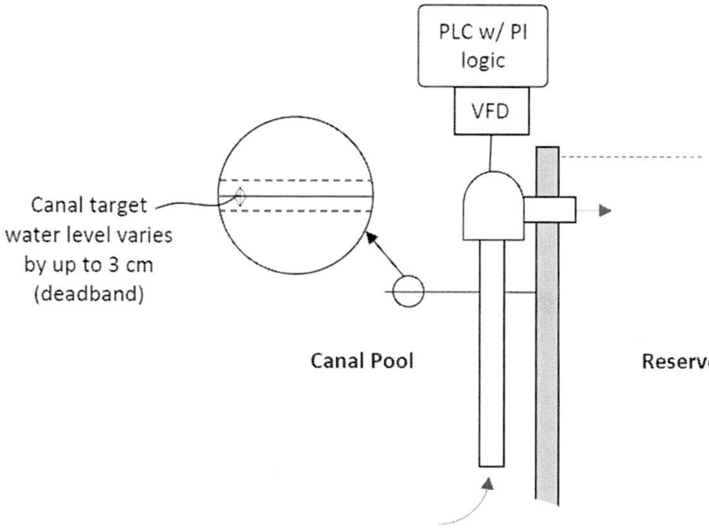

Fig. 2-58. Pump control option A

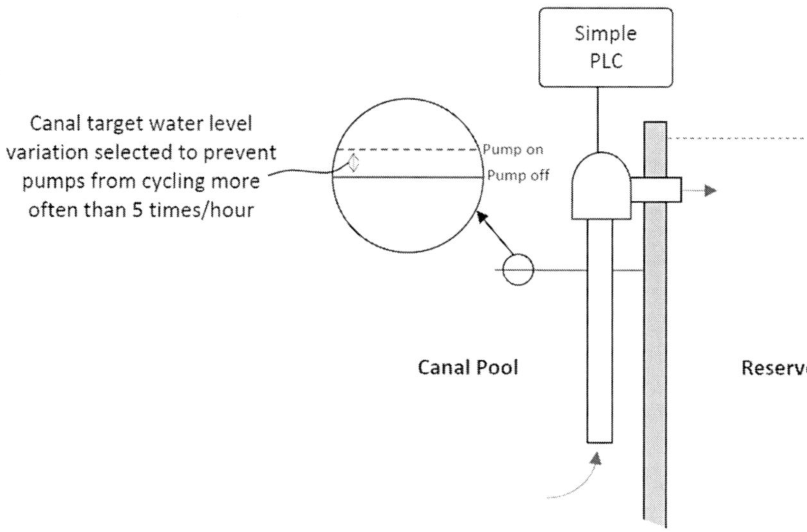

Fig. 2-59. Pump control option B

Option C: Probe contact sensors (not transducers), no PLC, and single-speed pumps (Fig. 2-60). Each pump has a different "on" and "off" water level. This is an old control technique that results in large canal water level fluctuations but is simple to understand.

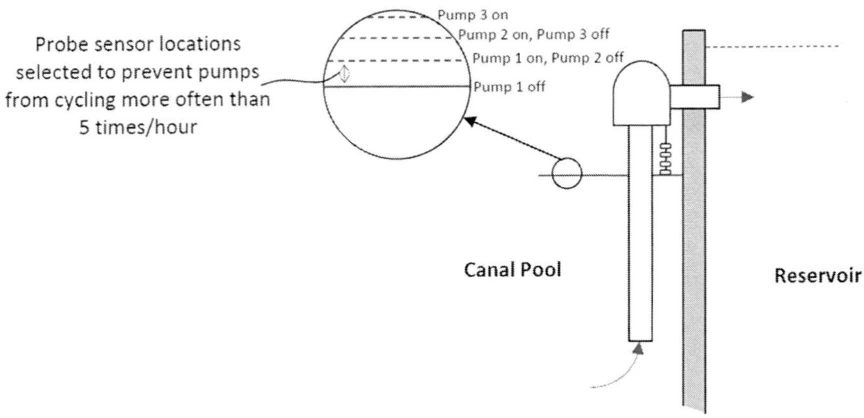

Fig. 2-60. Pump control option C

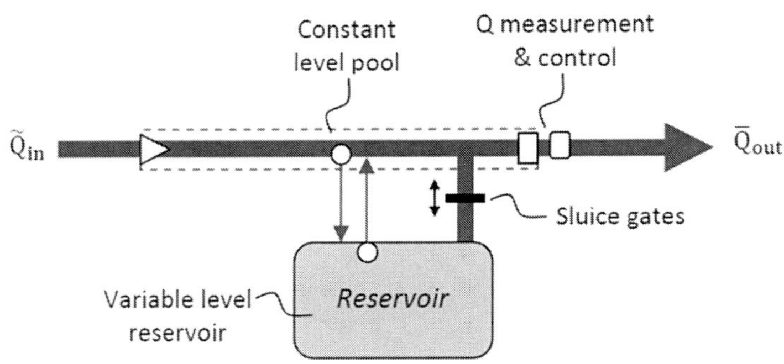

Fig. 2-61. Gravity in and out, and pump in and out on one reservoir

Control of Flows out of a Reservoir Water exits the reservoir to prevent the water level in the adjacent canal pool from dropping below its target. As mentioned earlier, there are two basic control designs:

1. If the water level in the reservoir is at or below the water level in the canal, pump(s) must be used. These can use one of the three pump control options described earlier.
2. If the water level in the reservoir is at least 8 cm higher than the water level in the canal pool, the flow can be pulled by gravity (using PLC-controlled gates).

Sometimes, as in Fig. 2-61, there may be gravity in or out and pump into or out on the same reservoir, depending on the water level in the reservoir.

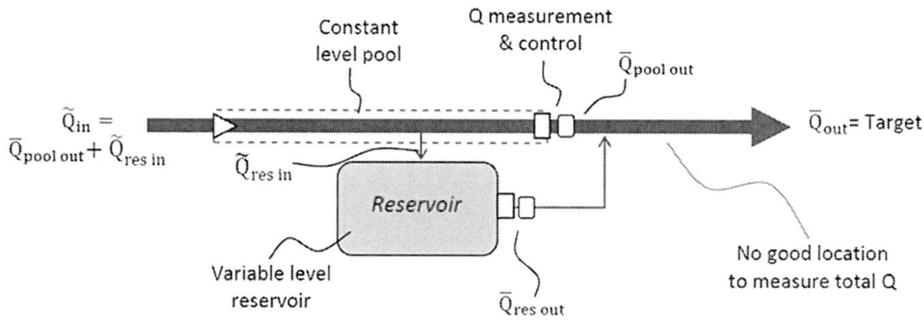

Fig. 2-62. Location with two flow control and measurement sites

Gravity Outflow When there is a drop, allowing gravity outflow from the reservoir, there are important details regarding how the flow rate out of the reservoir is computed and controlled. Two common situations are described:

1. There is no good location, or conditions are poor, for an in-channel flow measurement device downstream of the reservoir outlet on the canal. For example, there may not be enough head difference to install a broad-crested weir. Or there may be no good straight section available to install any flow measurement device.

   In this case, there must be two excellent flow control and measurement sites (Fig. 2-62). Both $\bar{Q}_{\text{pool out}}$ and $\bar{Q}_{\text{res out}}$ will have a set flow rate, with the sum of those flows equal to the canal's target flow rate (T_b). Note that for this control scheme, $\bar{Q}_{\text{res out}}$ must be accurately measured.
2. There is a good location to measure the total channel flow rate in the canal pool downstream of the reservoir discharge point. In this case, the two flow control points do not need to have accurate flow rate measurement. The flow rate through the large canal gate is constant (because it has a constant head on it), but it can be adjusted manually. The gate from the reservoir typically must be automated to adjust for varying reservoir water levels. This reservoir discharge gate will adjust its position automatically to maintain a total target flow across the flow measurement device.

   One advantage of this option is that the "total channel flow rate" can be measured downstream of a section with numerous turnouts. The actual discharge from the reservoir plus canal will equal the target flow rate, plus the turnout flow rates. The turnout flow rates can be varied with great flexibility, and the reservoir discharge gate will compensate automatically (Fig. 2-63).

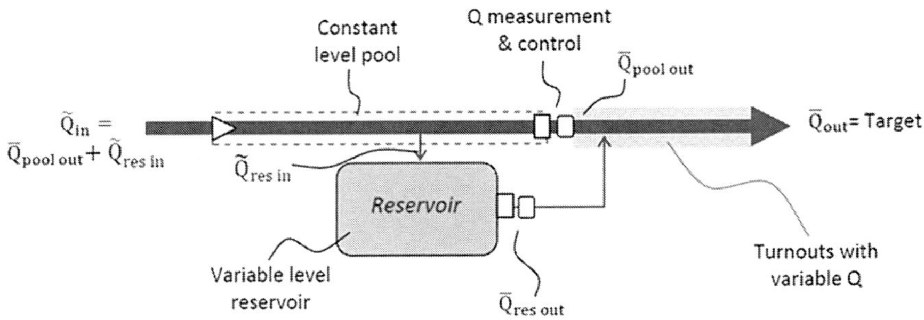

Fig. 2-63. Example with two flow control points

2.7. REFERENCES

Belaud, G., Litrico, X., de Graaff, B., and Baume, J.-P. (2008). "Hydraulic modeling of an automatic upstream water level control gate for submerged flow conditions." *J. Irrig. Drain. Eng.*, 134(3), 315–326.

Howes, D. J., Burt, C. M., and Sanders, B. F. (2010). "Subcritical contraction for improved open-channel flow measurement accuracy with an upward-looking ADVM." *J. Irrig. Drain. Eng.*, 136(9), 617–626.

Litrico, X., Belaud, G., Baume, J-P., and Ribot-Bruno, J. (2005). "Hydraulic modeling of an automatic upstream water-level control gate." *J. Irrig. Drain. Eng.*, 131(2), 176–189.

CHAPTER 3
SCADA SYSTEMS

3.1. INTRODUCTION

Supervisory control and data acquisition (SCADA) refers to a broad and ever-changing spectrum of electronic hardware, computer software, and communications infrastructure that provides a platform for remote monitoring and control in a variety of industrial applications. In canal applications, SCADA system complexity ranges from simple systems that allow operators to monitor a few water levels or flow rates over a radio network to large-scale, multiserver systems capable of automatically controlling large canal networks over fiber optic and microwave communication networks. Whether small or large, SCADA systems can provide real-time monitoring, remote supervisory or automatic control, troubleshooting, and automatic data reporting and archiving capabilities.

Until the 1980s, SCADA equipment consisted of expensive software running on large, expensive, dedicated computers in a central location. These computers communicated with remote sites outfitted with field equipment possessing functionality that is considered limited by today's standards. Today, a basic but complete SCADA base unit can be built entirely within a standard desktop computer at a relatively low expense. At remote sites, technological advances now allow for small, fairly inconspicuous, and inexpensive solar-powered installations with enhanced capabilities that allow one to monitor water levels or control gates, pumps, and other equipment. One beauty of today's SCADA technologies is that generally they are scalable, meaning that through some long-term planning and proper component selection, it is possible to start with a small implementation that can be expanded by adding functionality with minimal infrastructure changes.

Whereas some irrigation districts may have in-house expertise capable of implementing a SCADA project, others may need to rely on professionals who specialize in the development and implementation of industrial control systems, generally referred to as *system integrators*. Because of the underlying complexity, cost, and other factors, the adoption of SCADA systems in irrigation and drainage was relatively slow, and few integrators had experience in the realm of irrigation and drainage. Technological advances, decreased hardware and software cost, and increased governmental regulation have resulted in more interest in the application of SCADA for canal operations. With this increased interest, more system integrators are developing expertise in the implementation and maintenance of SCADA systems for canal management. In turn, with increased experience, the days of custom implementations, it is hoped, will give way to multiple integrators offering off-the-shelf solutions for canals.

The idea of standard practices is mentioned multiple times in this chapter. Implementation time and especially maintenance overhead can be greatly reduced by establishing standards for wiring, equipment placement, database configuration, naming conventions, and operator interfaces, among other things. Redundancy also is addressed in a few areas of SCADA. Although some may see certain redundant items as a luxury, redundancy can lead to more reliable control, improved customer service, greater integrity of overall operation, and security of historical data.

Like many technical domains, the SCADA world is littered with acronyms and abbreviations. Many of these shortcuts are standard throughout the industry, but manufacturers, sales staff, system integrators, and district employees have their own definitions for a few of the commonly used SCADA acronyms, and some of these definitions continue to change over time.

There is also technical terminology that commonly is applied in varying contexts. *Client–server* is one such overloaded term; it is used as a descriptor in multiple facets of SCADA ranging from communication models to hardware and software implementations. As used in this chapter, *client* refers to entities making requests for operations to be performed, and *server* refers to entities performing the operations. A common example is a print server on a computer network. Client software on one or more workstations instructs the print server to carry out printing tasks. The server performs the tasks and reports the success or failure back to the client.

Some of the technologies that make up SCADA tend to advance quickly. Any reference to specific hardware and software technologies, manufacturers, or communications protocols is meant solely as an example and should not be considered an endorsement or recommendation. One of the most quickly changing arenas is the use of wireless mobile electronics. Twenty years ago, we saw handheld touch-screen tablet devices in science fiction television and movies. Today, they are commonplace, and their utility is increasing daily. Use of these devices in SCADA is still in its infancy, and there is great opportunity to expand their use.

For some, examining the details of a new SCADA implementation or the modification of an existing system can be overwhelming. There are a number of decisions to be made in the development of a SCADA system. These decisions might include whether to use in-house expertise or hire an integrator, whether to choose a proprietary or open system, whether redundant sensors are in order, and what choice of communications protocol is appropriate. The purpose of this chapter is not to answer these types of questions but rather to give some useful basic information for consideration of such issues.

3.2. BASIC SYSTEM COMPONENTS AND FUNCTION

In a broad sense, SCADA systems are quite simple. The word "supervisory" implies that there is some human oversight, typically at a central base unit through a human–machine interface, or HMI. The HMI provides facilities for data storage, manipulation, and visualization. The base unit communicates with one or more remote units through some communications infrastructure, reading information, and issuing control instructions generated either by the human operator or automatically through software in the base (Fig. 3-1). The remote unit communicates with sensors, implements control instructions issued from the base unit, and if desired, the remote unit can implement local control functions programmed in the remote hardware itself.

3.2.1. Digital Communication

Uninterrupted communication is the backbone of a SCADA implementation. The system essentially is useless when the communication system is not functioning.

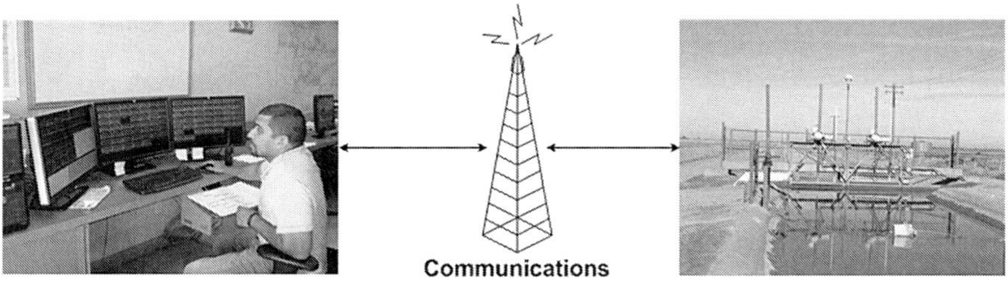

Fig. 3-1. Basic SCADA components

In the current state of technology, the hardware at both the remote site and base usually is quite capable of handling the needs of the typical canal implementation in terms of computational speed. The factor limiting the effectiveness of a SCADA system is more likely to be the throughput and availability of information through the communications infrastructure.

Outside of the signals exchanged with analog sensors and actuators, today's electronics are almost completely dependent on the use of digital information and digital communication. Although we think of canal information such as water levels in terms of analog values (e.g., a 2.5-m water depth), the hardware must convert this information to a series of 1s and 0s, called "bits" in order to make use of the information internally and to communicate the information to other entities.

Communication Models There are a number of models that describe how entities communicate with each another. The *master–slave* communication model has been a historical standard in SCADA. The term has come under cultural scrutiny in recent years and will be referred to here as *client–server*, as mentioned previously. In this model, the client drives the communication with one or more servers. Typically the base unit acts as the client and remote units act as servers (Fig. 3-2). The client issues commands to each server, typically asking for information or writing values to the server, and the server provides a response.

There is no direct communication between server units. Thus, if one server needs a value from another, the client first reads the value from the appropriate server and then writes it to the other. The ability for a server to self-report to the client can be an appealing communication component, especially for alarm or security reporting by the remote unit. The ability to self-report is not guaranteed in client–server configurations due to deficiencies in components such as the communications protocol or the physical communications link.

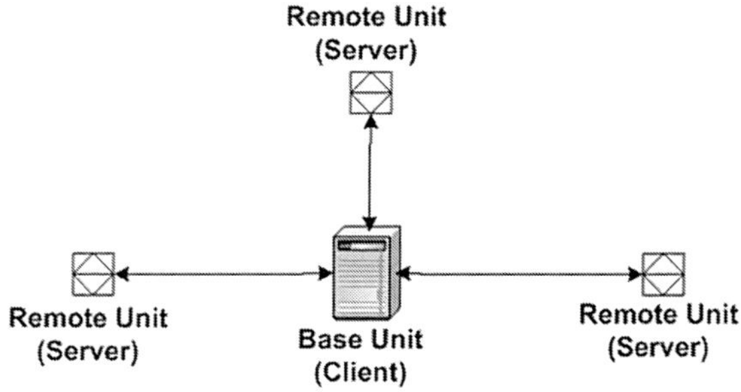

Fig. 3-2. Client–server model

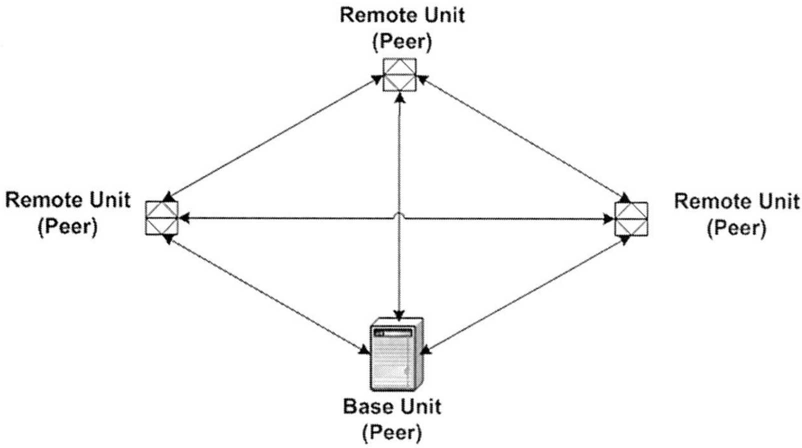

Fig. 3-3. Peer-to-peer model

In the *peer-to-peer* model (P2P), all entities can address each other directly (Fig. 3-3). In this case, the base unit does not have to act as the middleman if information needs to be sent directly between remote sites. Self-reporting is implicit in the P2P model.

The differences in these two models typically are related to the manner in which the communications protocol implements unit addresses. Additionally, certain physical communication configurations may not allow two entities to be party to the same communication stream. So even though a P2P protocol has been implemented, systemwide P2P communication is not guaranteed.

Communication Network Topology Network topologies are helpful in visualizing the function of a communications system from a number of perspectives. The simplest topology is *point-to-point*, which connects only two entities.

A *bus* topology consists of multiple addressable entities connected to the same communication stream, for example, through a common cable or over a shared radio frequency (Fig. 3-4). In the bus topology, all entities are party to the full communication stream. Any entity may transmit, and all receive a given signal. A given entity only processes incoming communication directed to its designated address. An example of a bus network would be a set of remote units connected to a base unit over a serial radio network.

In a *star* topology, each addressable entity is connected to a central switch in a point-to-point fashion (Fig. 3-5). All communication must go through this central point, and the switch directs the communication

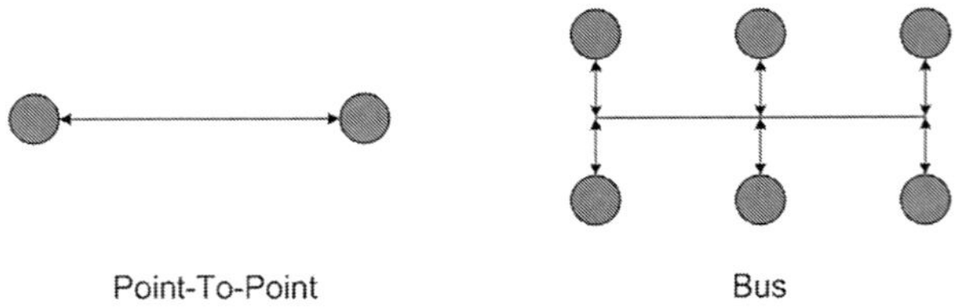

Fig. 3-4. Point-to-point and bus topologies

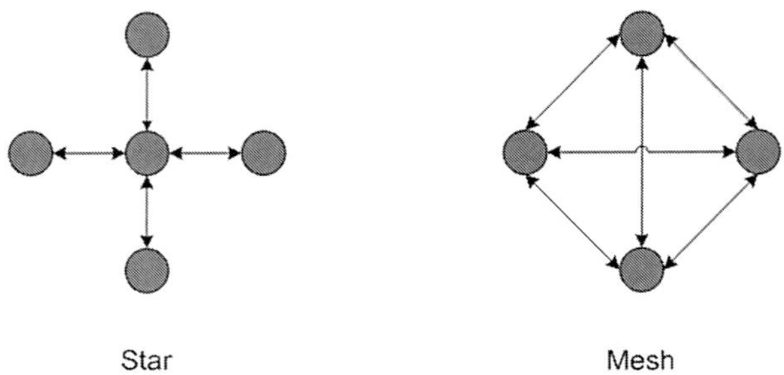

Fig. 3-5. Star and mesh topologies

appropriately, based on addressing. A typical wired office computer network managed through a network switch is an example of a star topology.

A *mesh* network topology allows direct interconnection between multiple entities. In a full mesh network, all entities are interconnected. Partial mesh topologies are also possible. There is a variety of mesh network implementations and there is significant detail in the mechanisms that govern the movement of information through such networks. Putting that detail aside, the basic idea in a mesh topology is that information moves around in the network until it reaches its destination. One advantage of mesh network topologies is that these allow for self-healing infrastructure. Should one node in the network go out of service, the remaining nodes can reconfigure the topology and continue to operate. When the missing node becomes available, the mesh then reconfigures to include that node.

These topologies can be combined. Consider a base unit connected to some remote units on a serial radio network, as well as to a distant set of remote units, by way of a point-to-point microwave link (Fig. 3-6). From the functional perspective of the actual communications stream, this arrangement could be seen as a combination of a bus topology with a point-to-point/bus extension. If the point-to-point link were to stop working, there could be communication between the distant remote units,

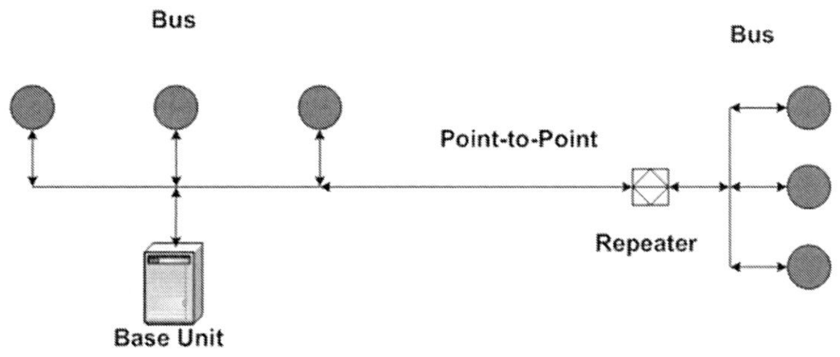

Fig. 3-6. Combination of bus and point-to-point topologies

but there would be no communication between the base and remote units serviced by the microwave link.

Modes of Communication Operation *Simplex* communication describes communication in a single direction. A common example is commercial radio. The radio station transmits a signal but does not receive anything. This mode would be applicable to a monitoring-only SCADA system in which the remote locations simply self-report.

Bidirectional communication is termed *full-duplex* when the infrastructure allows a device to send and receive information at the same time. Some combinations of hardware and communications protocol can offer full-duplex communication. An advantage of full-duplex communication is improved throughput. However, a disadvantage of full-duplex communication is that troubleshooting full-duplex communication can be difficult. An example of full-duplex communication is the use of the Internet for a teleconference. Colleagues in France, Australia, and the United States can see and hear each other in real time. Information "packets" containing audio and video information are transmitted and received simultaneously at each location.

Half-duplex communication only allows one party to talk at a time. In a client–server protocol architecture, once a client sends a signal, the channel remains silent until either the specified server responds or a response timeout occurs. Although this downtime may not be critical in a strictly supervisory system with a slow polling rate, it can become quite important for large systems or systems implementing centralized control.

Half- versus full-duplex has implications in information throughput limits, as well as various design decisions, including protocol selection, selection of transmission medium (e.g., radio, fiber optic, cell phone), and in the case of fixed-frequency radio, the number of radio frequencies required to implement a given communications configuration.

Communication Processing Various industry groups and manufacturers have established standards and protocols for both hardware and software that specify the format and transmission of information in a digital format between stations. The International Organization for Standardization (ISO) has established the open systems interconnection (OSI) model to characterize these standards into seven layers. This model is useful for visualization and planning. Basic familiarity with the layers and their interaction can be helpful when selecting hardware and software for both the base and remote stations, as well as selection of communication infrastructure.

As shown in Fig. 3-7, communication passes from the application layer of one device, through the layer stack to the physical layer, and then retraces the path on the receiving end. As the communication descends through the layers, additional information is attached to one or both ends of the basic data packet to fulfill the requirements of that layer. Then, as the communication ascends the stack on the other end, this extra information is stripped off, eventually transporting the data packet to the receiver.

Some layers are unused in certain protocol and hardware combinations. As an example, consider the Modbus RTU (remote terminal unit) serial protocol (Fig. 3-8). Modbus RTU was developed by Modicon (now Schneider Electric) for use with its equipment, and it has been a de facto

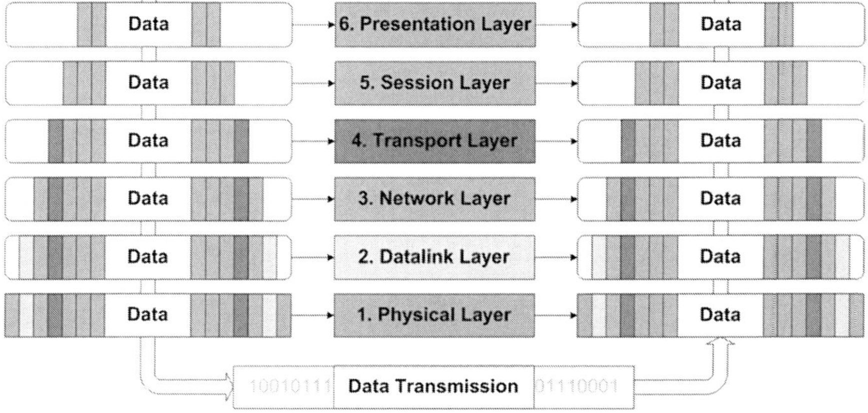

Fig. 3-7. The OSI communication model

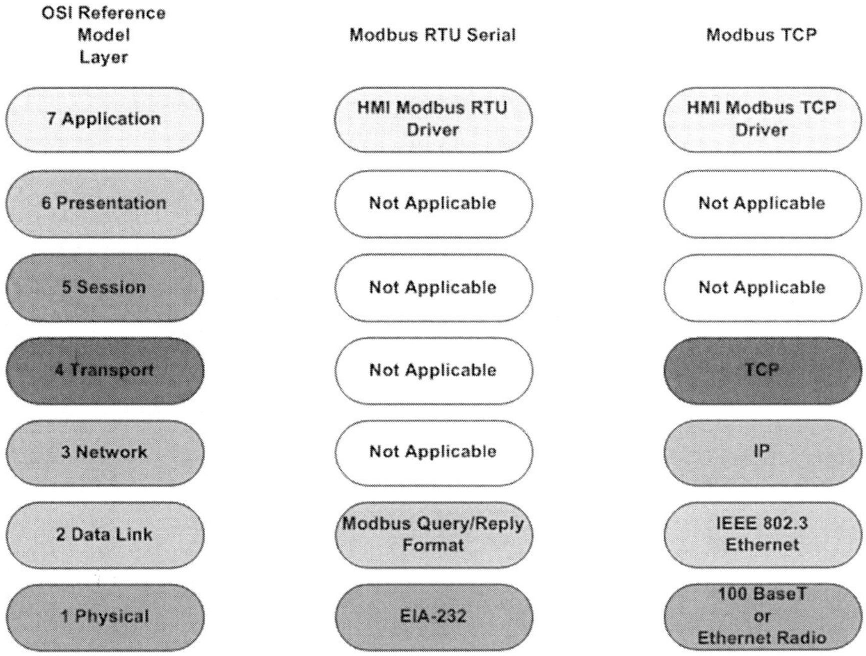

Fig. 3-8. The OSI model applied to Modbus RTU and Modbus TCP

standard for decades. A Modbus message is initiated by the Modbus communications driver at the application level. Levels six through three are not applicable in Modbus RTU. The Modbus query and reply format is applied at the data link level. At the physical layer, the information is sent via a serial interface (e.g., EIA-232 or 485).

A serial Modbus RTU implementation can transition into a networked Modbus TCP system. A Modbus TCP driver is implemented in the application layer. Transport control protocol (TCP) and Internet protocol (IP) are added into the stack to handle the switching and peer-to-peer aspects of the networking, and Ethernet replaces the query and reply format. Finally, the serial line is replaced by a network compliant standard, such as 100Base-T or Ethernet radio, at the physical layer.

Although the implementation of this transition from RTU to TCP is far from being simple, it serves as an example of how appropriate component selection can lead to a flexible long-term solution. Typically, HMI products and field hardware allow the use of a number of protocols on a variety of communication ports. In the previous example, if the HMI and the field hardware each support both versions of Modbus and each have a serial port and an Ethernet port, the transition requires software modifications and replacing the serial connection with an Ethernet connection (or the radio equivalent). Note that the transition could be between any protocols. Modbus simply provides a simple example demonstrating the utility of viewing protocol implementations through the OSI model.

Many HMI software packages allow multiple drivers to run simultaneously, such as Modbus RTU over serial radio and Distributed Network Protocol version 3 (DNP3) over Ethernet radio. The same is true for many remote hardware systems as well. This arrangement has a number of advantages. First, it allows migration from one protocol or hardware implementation to another over time. This can be helpful for a district considering modernization on top of an existing physical communication infrastructure, as it does not require an immediate, full-scale transition. Second, the ability to support multiple communication streams simultaneously allows for the implementation of redundant communication. Finally, the ability for remote site hardware to handle multiple protocols in varying roles can provide enhanced local communication flexibility at the remote site. This will be discussed in detail in the remote unit section.

Numeric Format As mentioned earlier, digital devices utilize binary information. Generally, this translates internally into integer values (i.e., –1, 0, 1, 2, …). We think of measurements in terms of real numbers, again, the example of a 2.5-m water depth. Standards have been developed to represent real values in a binary format and to establish the binary arithmetic operations used for calculations. For common applications, the format is referred to as *floating point*. Typical personal computer architectures use an 80-bit format to represent floating-point values, whereas some floating-point capable field hardware may use a 64- or 32-bit format. Generally, floating-point differences are transparent to computer users. When a real value is typed into a spreadsheet cell, the computer codes it into the closest 80-bit binary floating-point value. However, with differing coding and display formats between the base and remote systems, it is possible to observe one number while standing at the field unit but see a slightly different value displayed at the base.

Another numerical stumbling point can occur due to the order in which varying devices use, store, and transmit floating-point information. Known as *endianness*, this order can differ between devices. Typically, communications protocol drivers are capable of handling the various ordering combinations. It is a matter of making sure both entities on either end of the communication are matched. Otherwise, the information will not be interpreted properly.

Communication Error Detection One necessity in any communications protocol is some sort of method for error detection. This is typically implemented as some type of check sum that is computed from the transmitted bits and attached to the end of the message. The receiver computes its own checksum as it receives the message. If the checksums do not match, the message is ignored and typically some sort of error is logged. Without some type of error detection, random noise could be interpreted as a valid signal, possibly resulting in catastrophic consequences.

Communications Protocols Protocol selection is important in SCADA design and implementation, as it affects hardware and software selection, as well as information throughput, communication mode, and communication topology. Protocols can be lumped into two basic groups: *open* and *proprietary*. Typically, open protocols allow for the selection of hardware and software from a variety of vendors. Whereas the basic mechanics are defined by the protocol, the open nature of the protocol can lead to some hybrid implementation schemes.

Modbus RTU already has been discussed as an example of an open protocol. Modbus RTU is a simple serial protocol that implements the reading and writing of a basic set of data constructs utilizing a client–server format between a single client unit and multiple server units. Modbus gained an early foothold due to its simplicity and is still supported today. It is so simple that with some experience it is possible to interpret the Modbus RTU communication stream in its raw numeric format by simply capturing and displaying the stream in real time. With this simplicity come a few drawbacks. The throughput implications of the client–server format are the first drawback. Second, the small set of data constructs can be limiting. For instance, Modbus RTU has no defined format for floating-point values. This is an instance where endianness can come into play. Finally, there is no data quality information included in the Modbus RTU communication.

The distributed network protocol (DNP) is an open protocol developed by the electrical utility industry. Now in its third major revision, DNP3 is essentially a point-to-point protocol. Manufacturers have utilized formatting within the DNP3 data structures to implement peer-to-peer support. Although the DNP3 implementation is not as simple as Modbus, DNP3 has seen increased use in the past decade in the oil, gas, and water industries.

A number of industrial control companies cooperated with Microsoft to leverage object linking and embedding (OLE) technologies to implement interoperability between devices in distributed systems. The result was OLE for process control (OPC). Today, OPC is administered by the OPC Foundation and is supported by many hardware and software systems. OPC offers infrastructure for data exchange, alarms, and access to historical data.

Proprietary communications protocols are offered by many hardware manufacturers. Although using a proprietary protocol might limit the selection of components, these protocols can offer some appealing enhanced functionality. An example of such a protocol is the Motorola data link communication (MDLC) protocol. MDLC supports peer-to-peer operations, remote diagnostics and configuration, and encapsulation or "transcoding" of other protocol formats, robust data quality information, and encryption capabilities.

Radios Canal networks commonly service large geographic areas. Additionally, these areas can be remote and sparsely populated and, therefore, lack much in the way of communication infrastructure. For these reasons, radio is a popular communication medium for SCADA operations on canals. Relatively inexpensive and reliable radio systems typically are owned and controlled by the owner of the SCADA system. There are no monthly service charges, failure rates are better than telephone failure rates (there are no long wires to fail), and if repairs are necessary the owner controls the repairs (as opposed to repair priorities established by the phone company repair service).

Radio selection depends on many things, but most important to consider when selecting a radio system are the terrain and the distance the signal must carry. These two factors affect the radio frequency and the required transmitting power.

There are tradeoffs based on the operating frequency of the radio system. VHF (very high frequency) radios operate in the frequency range from 30 MHz to 300 MHz while the UHF (ultra high frequency) radio band extends from 300 MHz to 3 GHz. As a general rule, lower frequencies mean greater radio antenna size, greater transmission distance, greater chance of interference from electrical sources, and lower interference from topography, foliage, and buildings. Higher frequency systems offer greater isolation from electrical sources and smaller antennas, but topography, buildings, and plants are more apt to cause interference. Transmission distance also decreases at higher frequencies. In general, line-of-sight is best, but VHF bands are more forgiving.

In the United States and some other countries, the VHF and UHF radios can be either licensed or unlicensed. The main difference between licensed data radios and unlicensed data radios (*spread spectrum*) is the transmission power that is permitted. Spread spectrum radios can only transmit with 1 watt of power, while licensed radios can transmit up to 5 watts.

Both technologies have their benefits and drawbacks. Licenses may not be available in a given area for the frequencies that work best. This is particularly true in more densely populated areas. If licenses are available, application processing can take a long time. There are also limitations on how much data can be transmitted (*bandwidth*) with licensed radios. Licensed data radios are usually okay for SCADA systems where data are sent from monitoring stations to a central office (specified systems typically require 9600 baud, even for districts with many sites). However, if the district would like to use cameras or send significant amounts of data from a pumping plant with variable frequency drive (VFD) controlled motors, the data can overload a licensed system, and the speeds are too slow.

If a licensed radio system is selected, it is important to specify the use of two frequencies. The use of two frequencies allows the radio to use one to transmit and one to receive. It also allows for full-duplex communication, which is much faster and can be more reliable. It is possible for radios to use a single frequency for both transmitting and receiving. However, this mandates the use of a half-duplex infrastructure and slower communication. Single-frequency systems generally are limited to older systems.

Wireless Ethernet radios provide state-of-the-art, high-speed operation with the capability of transmitting real-time images or video. In addition, information is readily available to the local network in the office in a standard IP configuration. Potential applications include sending an image of the status of a canal water level and surrounding conditions after an alarm condition occurs. Another possible application is to take and send a picture from a security camera of people entering or exiting a facility. It is also possible to locate Wi-Fi access points at remote sites equipped with Ethernet radios. This can facilitate the use of wireless laptop and tablet computers for troubleshooting problems at the remote site.

A system of modern data radios in good operating condition should have error rates less than 2%. This means that more than 98% of the radio communication attempts are successful. Furthermore, if there is an error, the radio should automatically retry the transmission. There should never be communication gaps of more than a few seconds.

It is always recommended to have radio tests done before making a final selection of radio equipment. The radio test should be done with the actual equipment to be used in the system and should cover path surveys for every site. Vegetation can affect radio communication. The radio test should be conducted when deciduous trees are fully leafed out. Additionally, the test should accommodate tree growth over time. A good radio test guarantees that signal strengths will be at acceptable levels and transmission will occur at reasonable speeds.

Beyond Radios Whereas two-way radio VHF and UHF communication is highly utilized in canal SCADA implementations, there are other options. Microwave communication uses a focused, high-frequency signal capable of handling high throughput. Microwave is useful for long-distance point-to-point transmissions. Cell phone technologies such as 3G and 4G provide high data rates as well but can be susceptible to system outages, and there may be little or no service in remote areas. Satellite communication is another possibility, especially for very remote locations. Generally, these support lower data rates. High data rates are possible but can be expensive.

Communications Security In the past decade, security of vital infrastructure has come to the forefront. The level of security required is quite variable and depends on a number of factors, including location and government requirements.

SCADA communication systems can be vulnerable. Outside of protecting the physical communication infrastructure, consideration should be given to the security of the data stream itself. This is especially true in the case of radio communication. Ultimately, the information required to gain access to the data stream and record information or to carry out a malicious act is fairly complex. At minimum, one needs to know radio frequencies, the protocol, and information specific to the hardware at the remote unit. Technologies like frequency-hopping, spread-spectrum radios provide another level of protection. To gain access to this type of communication stream, the specific channels used in the frequency-hopping scheme, as well as the hopping order, need to be known. Data encryption adds yet another level of security. Varying levels of data encryption are available in protocol implementations, as well as communication hardware.

3.2.2. The Remote Unit

RTUs, PLCs, and PACs This is an area where SCADA abbreviations can become a bit muddled. As SCADA was developing, there were distinct differences between remote terminal units (RTUs) and programmable logic controllers (PLCs). Traditionally, RTUs were slow, used high-level programming, had high power consumption, and possessed a lot of computer memory. RTUs were used mainly for data acquisition. PLCs, conversely, were relatively fast, programmed in the rule-based language Ladder Logic, and consumed less power but had less memory. PLCs implemented deterministic control by running the control program in a single master loop at a defined interval.

In recent years, with advances in computer technology, the differences between PLCs and RTUs have almost disappeared. In fact, many are beginning to use the terms differently than their original definitions. Now, many call the "package" at a remote site an RTU. PLC is now frequently used to describe the industrial computer within this package. Using this nomenclature, the PLC is one of several components in an RTU.

Nonetheless, it is still common to see or hear the whole package referred to as a PLC.

The hardware world has evolved further with the development of the programmable automation controller (PAC). The PAC represents a blend of RTU ruggedness and connectivity with the function and programmability of a personal computer. This blend provides the ability to implement complex numeric routines within the PAC itself. However, this flexibility comes at a cost, as it is up to the programmer to ensure deterministic operation.

For the sake of simplicity, unless distinction is necessary or helpful, the intelligent control and data acquisition unit at a remote site (RTU, PLC, or PAC) is simply referred to as a PLC from this point forward.

Historically PLCs were available in defined packages. A given model had a fixed number and type of inputs for reading sensors, outputs for implementing control, and communication connections for interaction with the base unit, local operator interfaces, and other functions. Today, flexible, modular, and expandable units are available. Should the need arise for more analog input ports, it is possible to purchase a module and plug it into an open space on the controller backplane. However, it should be noted that with this flexibility there can be additional complexity in programming and configuration, as well as additional expense.

The PLC represents the intelligence at a SCADA remote site. It controls all the step-by-step actions at a site. PLCs receive the information from the field instrumentation, run programs, process local control actions, and transfer the appropriate information to the central SCADA station.

The PLC is an *embedded system*, meaning that the unit operates as a stand-alone, autonomous system. All of the software required for the unit to operate is maintained in the nonvolatile onboard memory. Unlike a personal computer that allows the user to add and remove multiple programs at will, the typical PLC implementation can be thought of as a well-defined set of routines that change only when necessary.

Many PLC manufacturers provide an interface to develop customized programs for a given implementation. Historically, PLCs were programmed using Ladder Logic. This language originally was developed to mimic the indicator lights, switches, and relays in older analog relay-based systems. Ladder Logic and a number of variants are still in use today. However, the advancements in processor capabilities have given way to programs written in standard text-based languages, such as C++, as well as graphical programming languages such as LabVIEW from National Instruments that represent the code as graphical objects.

The programming environments provide varying levels of resources for debugging code, including execution breakpoints, single-step execution, and examination of variable information during program execution.

In some development environments, it is possible to test PLC code within a software simulation of the hardware. This facilitates code development without the need to have the hardware constantly present.

Even with all of the amenities offered by the modern programming environments, development of PLC code can be a cumbersome task and in most instances may be best left to a professional integrator or programmer. As modern SCADA implementations become more commonplace on canal networks, it is conceivable that SCADA integrators will provide PLCs preprogrammed with software capable of providing support for various sensors and functionality typically used at remote canal sites.

Standardization is important in the implementation of custom PLC software. Significant time can be added to the task of debugging a problem at a remote site should the technician have to determine which software version is running at that location.

Other facets of standardization apply to the PLC code itself. It is advisable to refrain from using embedded constants in the code. Although it might be appropriate to hard-code a universal constant such as the acceleration due to gravity, generally values such as control, scaling, and counting parameters should be available for adjustment through communication with the base unit, as well as through a local interface.

Another aspect of that standardization is related not only to the programming but to the communications protocol as well. Many communications protocols are register-based, meaning that they read and write values, or groups of values in the PLC based on some defined addressing scheme. PLCs generally provide a mechanism to map internal values, such as a water depth, to a specific register address used by the communications protocol. Standardization of this mapping is important as it can simplify configuration and troubleshooting for both the base and remote units.

As mentioned previously, PLCs use binary information in their internal operation and to communicate with other microprocessor-based devices. The electronic module that actually reads analog sensors is called an *analog-to-digital* (A-D) *converter*. A-D converters typically are specified based on the number of bits used to represent the returned value. With a higher resolution A-D converter, a PLC can resolve the measurement from the sensor better because the range of the sensor output is "read" based on a finer scale. However, additional consideration should be given to the gains of using higher order conversion (14- or 16-bit) as this higher resolution A-D conversion is typically more expensive, and the increased resolution can be unwarranted, given some of the typical measurements in canal applications.

Many PLCs can be configured remotely. Updated programs can be transmitted over the communications system instead of going to each site to implement the update. This greatly eases the effort necessary to standardize PLC program versions and is recommended for large SCADA systems (i.e., ones with greater than 10 to 20 remote sites).

In addition to maintaining program code, remote configuration may also allow

- Maintenance of individual PLC configurations,
- Remote restart, and
- Remote diagnostics.

PLCs can offer user interaction through an operator interface terminal (OIT). An OIT can be useful for displaying system status, calibrating sensors, or operating actuators. The OIT can be an LCD display panel or a touch panel mounted in the door of the PLC enclosure. In areas prone to theft and vandalism, it is advisable to conceal the OIT. Another option is to connect directly to the PLC via a wired or wireless connection with a laptop computer or a wireless mobile device. To facilitate development of Web-based OIT technology, some PLC manufacturers have started including an onboard Web server, allowing the customer to assemble Web-driven operator screens for use by a local operator.

Local alarms are important for both the local operation of the site by an operator, as well as for transmission back to the base unit. Although these alarms generally are related to operational parameters such as water levels, flow rates, and gate positions (and their associated limits), they also can alert both field personnel and SCADA operators at the base unit to communication failures and intrusion or vandalism. Timely delivery of local alarm information through the SCADA communication network is governed by the ability of the remote unit to self-report. If a remote unit is not able to self-report, then the alarm information will not be available to the base unit operators until the next time the remote unit is polled for information.

Although most remote sites are not high-vibration environments, there may be some areas where vibration could be a concern. Temperature extremes also can be a concern at remote sites. In the interest of standardization, it is prudent to specify that PLC hardware should be of a suitable industrial quality to fulfill the needs of varied environments encountered throughout the implementation.

Many PLC manufacturers utilize optical isolation technology between each of the components of the system. This isolation provides a barrier to electrical surges in the system. Should lightning strike at or near a submerged sensor, it is possible that that surge could travel to the PLC. In a properly isolated system, the module connected to the sensor may be lost, but the optical isolation increases the chance that the remainder of the PLC components will not be harmed. It should be noted that this protection is dependent on proper isolation between the power supplies for each component.

Sensors Sensors in a SCADA system can be used to measure anything that can be converted into an electrical signal. In canals, sensors can measure water level, pressure, flow rate, water velocity, linear gate position (sluice gates), angular gate position (radial gates), pump status (on or off), pump speed, temperature, and more. Specific sensor technologies are discussed in Chapter 2. In general, there are at least three ways sensors communicate: analog signal, discrete signal, or digital signal.

An *analog sensor* communicates using either voltage or current. Using a predetermined range of voltage or current, the sensor measurement is directly proportional to the output voltage or amperage.

A *discrete sensor* communicates an on or off condition. A switch is either open or closed, a water sensor is wet or dry, and a motor is either on or off. Examples of devices using discrete signals are gate limit switches and motor relays. Many remote sites are isolated and prone to vandalism. A SCADA system also can provide security information through discrete sensors designed for intrusion and motion detection.

Intelligent or *smart sensors* are equipped with some computer intelligence. Typically they communicate through standard protocols discussed earlier, so they are addressable. Multiple smart sensors utilizing the same protocol may be attached to a PLC in a multidrop fashion such as a bus or star topology. This can be helpful when designing for sensor redundancy. Additionally, smart sensors can calculate output values internally and can communicate several different values from a single sensor. For example, some depth transducers also measure temperature. To deliver this additional information in an analog fashion would require additional conductors in the cabling between the sensor and the PLC, as well as requiring an additional analog input on the PLC. A digital depth sensor can send both the depth and temperature to a PLC in a single communication transaction.

Smart sensors give increased flexibility and can ease installation. Consider the case of a turnout located 200 m upstream from a regulating structure equipped with a PLC. Utilization of an analog meter would require either a dedicated PLC and radio or trenching and conduit to run a cable to the nearby PLC. A smart, solar-powered flow meter can be installed inconspicuously, tied directly into the primary SCADA communications infrastructure through an integrated radio, and provide additional information such as totalized volume.

Again, due to its simplicity, Modbus RTU has been a popular protocol in the initial implementation of some smart sensors. Modbus-based smart sensors still can be used in a SCADA system based on another primary protocol. Consider a remote flow control structure tied to the base unit

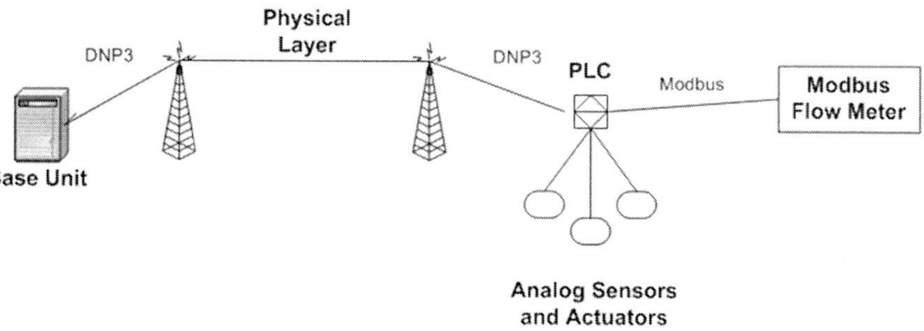

Fig. 3-9. Integrating a Modbus flow meter into a DNP3 infrastructure

through the DNP3 protocol over a radio network. As stated previously, many PLCs are capable of handling multiple protocols. If the PLC at the site is capable of functioning as a Modbus client, the smart flow meter can be connected to the PLC as a server unit (Fig. 3-9).

Whereas communication compatibilities may allow a smart sensor to operate as a stand-alone remote site, connecting smart sensors directly to the controlling PLC can be advantageous in local control situations. A local flow controller may require frequent flow measurements compared to the update rate required at the base unit. By connecting directly to the local hardware, traffic can be reduced on the main SCADA communications network. Flow information then can be passed from the remote site to the base over the main network as needed for operator updates and other functions.

Outputs and Actuators PLC outputs can be categorized into the same groupings as sensors. The use of analog outputs is limited as the circuitry for providing stable analog current or voltage output can be expensive. With the exception of some variable frequency drives (VFDs) for pumps, few devices currently in use on canals make use of such outputs.

Most actuators are directly driven by discrete outputs. Some PLC electronics are capable of switching higher AC voltages, but the current must be limited. For instance, the SCADAPack from Schneider Electric is rated for 240 V of AC at 5 amps. Other PLCs are limited to lower DC outputs. In this case, a relay is necessary when switching higher voltage loads and AC loads.

Similar to the sensor side, smart actuators also are available. In a manner of speaking, these devices have their own PLC that handles the positioning based on information sent from the primary PLC at the remote site.

Wiring All wiring external to the PLC panel should be installed in conduit and routed so that no part of the wiring is exposed to potential damage. To ease servicing and maintenance, all incoming and outgoing wires should be connected to numbered terminal blocks, neatly tied, and fastened to the chassis as required. Standardized wiring color and numbering is also very helpful. Shielded cable may be necessary to reduce noise in communication and sensor signals. The use of twisted-pair cable is also helpful for noise reduction.

All antenna leads, communications cables, power leads, and sensor leads should be protected from voltage or current surges caused by lightning strikes on or near the PLC, or on the water in the canal, reservoir, or pipeline. Lightning arrestors generally are specified by the type of cable or the equipment being protected.

In addition to the installation of lightning protectors, grounding is critical to successful lightning protection. Lightning protectors shunt excess voltage or current to ground. If there is not adequate grounding, the lightning protection system will not function properly.

Remote Unit Security Aside from the obvious physical security required to prevent intrusion or vandalism, another element of remote unit security is the specific configuration information for the PLC at the remote unit. Details related to the communication protocol addressing, as well as the register layouts used in the remote unit PLCs, could be used to compromise the system. Secure storage of this information should be considered.

3.2.3 The Base Unit

There are myriad possible configuration options for the base unit. In the most basic form, the base unit can be a single computer. For larger implementations, the base unit could be implemented as a set of networked operator consoles interacting with stand-alone database servers and multiple communications hubs. No matter the scale, certain basic components are generally common to all implementations.

The Human–Machine Interface

The human–machine interface (HMI) is the primary software component of the base unit. Although there are companies whose sole business is the development of HMI and accompanying software, some PLC manufacturers also offer HMI products. These proprietary HMI implementations may focus on communicating primarily with proprietary equipment at the remote end, but most allow the option of interfacing with equipment from other manufacturers, as well as with equipment compliant with various open communications protocols. The breadth of this external support varies from one manufacturer to another. There are three main components of the HMI: communications, database, and operator terminals. All three may be implemented on a single machine or distributed among multiple machines or even multiple facilities. Additional components include components for alarms, historical databases, report generation, and security.

HMI Communications The communications component of the HMI corresponds to the application layer (layer seven) of the OSI communication model discussed earlier. As previously stated, most HMI implementations allow the use of a variety of protocols that typically run as separate communication drivers. These drivers have user interfaces that allow technicians to configure the driver behavior and define the makeup of the information to be interchanged with each remote site. This configuration typically requires duplication of the PLC register mapping so that the base knows the proper addressing of PLC information. If the mapping at the base does not match that of the PLC, the data will be either mismatched (e.g., gate positions interpreted as water levels) or the mapping simply will be corrupt due to improper formatting (e.g., a floating-point value being converted into integers). Standardization of this mapping as well as the naming conventions applied to remote site and communication data elements can speed up the development of the HMI communications component and ease the effort in troubleshooting communications problems.

Database The database typically acts as the middleman between the communication drivers and the operator terminals by maintaining a snapshot of the current state of the system. The communication drivers populate the database with information coming in from the remote units. The database manipulates incoming information as necessary, stores that information as part of the current state, and updates the operator displays. In the other direction, any control actions specified at the operator console are stored in the database, manipulated, and then sent out to the remote units through the communication drivers. Some HMI packages provide for redundant database configurations. This can be helpful for organizations that reserve a backup base location as a fallback in case of a catastrophic event.

Granted, the "middleman" description as used is simplistic; SCADA database development, configuration, and maintenance can be a very involved and complex undertaking. Even for a simple supervisory system with no automatic controls, the database information could include alarming thresholds and sensor calibrations for each site, coefficients for computing flow through gates or over measurement weirs, and physical dimensions for structures such as spillways. Standardized naming and configuration conventions play an important role in the development and maintenance of the database.

Given the importance of the database in the overall function of the SCADA system, database availability must be ensured. To guard against hard drive failures, it is advisable to utilize a redundant storage architecture known as redundant array of inexpensive drives (RAID). RAID systems utilize multiple drives in varying configurations to provide redundancy. These hard drives can be mounted in hot-swappable mounts, allowing an inoperable drive to be powered down and removed without shutting down the entire system. Once a new drive is in place, the RAID technology provides mechanisms to repopulate the information on the new drive.

Routine backups of the database are highly recommended. Although the database can be complex, the actual size of the files comprising the real-time image of the system is comparatively small in today's information technology (IT) environment.

Operator Terminal The operator terminal provides the primary user interaction at the base unit (Fig. 3-10). Operator interface screens are usually developed from a basic set of tools for drawing and annotation, as well as sets of predefined on-screen controls such as push buttons, numerical displays, and trending graphics. Generally, the HMI provides the capability of programming these controls to customize their function.

Operators need to be comfortable with the operation of the system, both in normal day-to-day operations and in varying levels of crisis mode. When possible, it is important to consult the system operators when designing the screens that make up the interactive experience of the HMI.

Whereas an operator's primary focus is on managing the SCADA operations, most office environments also require the operator to run other management applications, as well as respond to e-mails and other routine tasks. Multihead video adapters have become commonplace and generally are supported by computer operating systems. These adapters enable the use of multiple display monitors for a single operator workstation (Fig. 3-11). The added display area allows the operator to control the system and operate other office software without having to switch to another computer completely.

The product licensing schemes for HMI packages typically allow multiple operator consoles to access the database and communications drivers, thereby allowing multiple operators simultaneous access to the system. Generally, this can include remote access from outside of the base unit. This arrangement can be advantageous in a number of scenarios. First, if the base is not staffed through the night, remote access enables an on-call

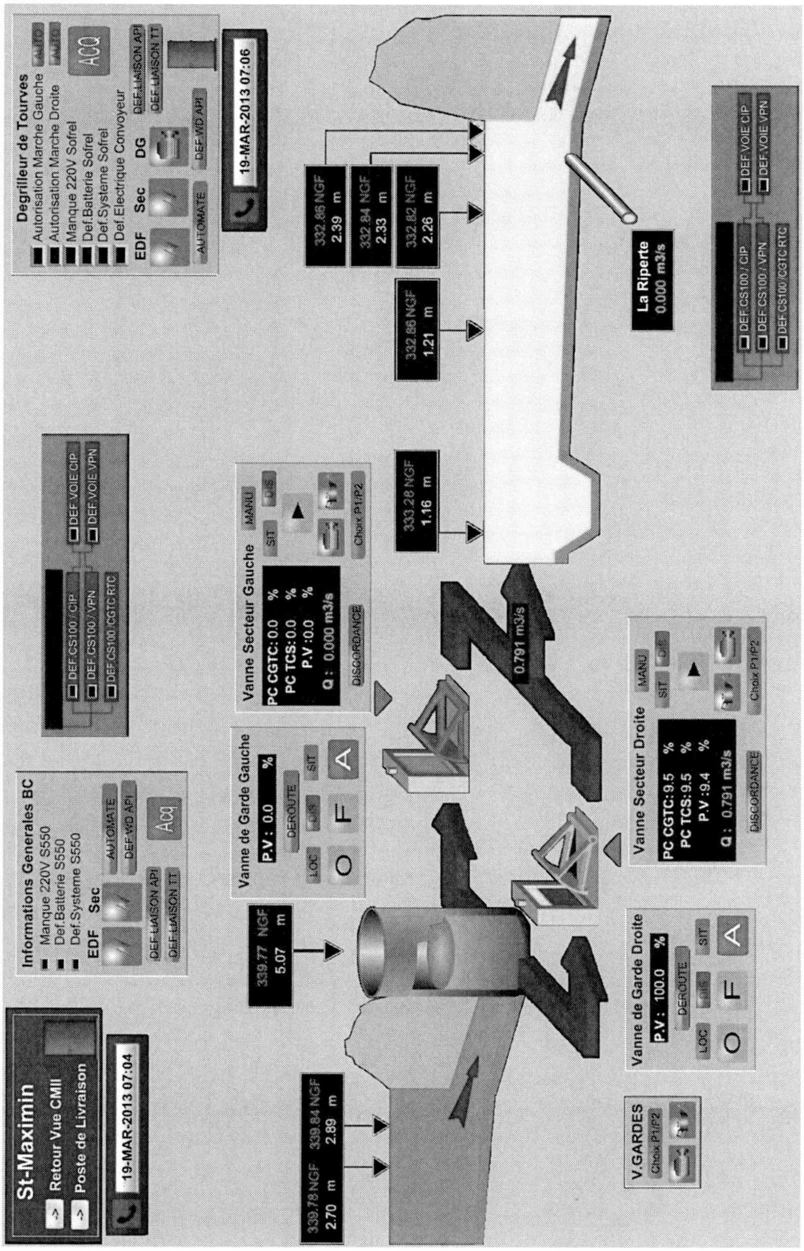

Fig. 3-10. Typical SCADA operator screen
Source: Courtesy of Société du Canal de Provence, Aix-en-Provence, France; reproduced with permission

Fig. 3-11. Operator workstation
Source: Photo courtesy of Central Arizona Project, Phoenix, Arizona; reproduced with permission

operator to respond to emergency situations from home through an Internet connection or even through a cellular connection. In this case the access would allow full control of the system. In another example, consider a manager who is offsite and wants to check the system status. This case requires only read access to the system. Access modes can be set through the HMI configuration or through the computer network management of the base unit.

Alarms Alarms are a focal point of base unit operation. Typically alarms are presented to the operator through visual indicators on the screen, as well as audible indicators for high-priority items. Depending on the brand of HMI, alarming functionality can be managed in the database, or it can be implemented in a separate process within the base computer architecture. One source of alarms is from the database. The database infrastructure generally allows developers to set the boundary values for "normal" operation. The developer typically is allowed to establish the boundaries of alarm conditions for varying levels of severity (e.g., high, moderate, or low). Should the database receive a value from a field unit or calculate a value internally that exceeds the bounds of normal operation, an alarm is communicated to a number of destinations. Typically they are displayed to the operator, many times with a repeating tone or alarm horn.

Alarms may also be routed to other devices, such as a printer, sent out as an e-mail, or directed to telephones either as a computer-generated voice message or as a text message. The mechanisms for this routing generally are provided either through database mechanisms, or through external third-party software.

Another source of alarms is the communication infrastructure. Should a remote unit stop responding, appropriate personnel, including managers, SCADA technicians, and even operators can be advised of the situations.

Historical Database Although the HMI database stores a current snapshot of the system status, it is also important to store historical information. This information can be incorporated into the operator displays, where it can be very helpful for diagnostics and can also be important for legal proceedings. HMIs generally provide historical databases optimized for storing time-based information. Instead of storing every value at a given interval, many of these databases only store values that have changed beyond some relative threshold within a given timeframe. The historical databases provide mechanisms to archive older data to offline storage, and many also offer the ability to encrypt the database information to prevent data tampering.

Reports HMIs offer a variety of report generation options. Developers can provide report templates that can be populated based on a timeframe specified through a user interface. In addition, many HMI packages offer the ability to export data in varying formats for use in spreadsheets and other management software.

HMI Security HMI software typically implements varying levels of user security by either granting or denying the ability to change values, view information, or issue control actions based on user or group designations. Some HMIs allow the implementation of security from within the development environment by providing infrastructure to designate user names, passwords, user groupings, and to prompt users for login information. Other packages base the security on the framework supplied by the operating system such as the security offered in Microsoft Windows. This has some advantages when the operator terminal is a multipurpose station acting as the operator portal for other day-to-day business aspects of the organization. By requiring each operator to log into the station through the operating system, user preferences and system settings such as e-mail login information are tailored to each user.

Additional Base Unit Software

Although the HMI might make up the bulk of the operator's interaction with the base unit, there are additional components that can be included in the suite of software used at the base unit. At many districts, base unit operators have multiple responsibilities, including taking delivery orders and collecting money. These functions can be facilitated through additional software packages.

Demand management software allows the entry of customer order information. In some cases this is done strictly by district personnel taking orders. Whereas some districts rely on telephone orders from customers, others allow field operators to accumulate orders directly from irrigators or farm managers. Today, field operators can be equipped with laptop and tablet computers for this task. Typically the orders could be downloaded upon return to the office at the end of the day. However, with the appropriate communication infrastructure and Wi-Fi enabled technology, updated orders can be downloaded automatically to the base unit whenever the operator is within signal range. Other demand management software allows customers to enter orders over the Internet. After order processing is complete, the customer can receive a confirmation that the order was accepted or a notice that it has been scheduled for another time through various means, including e-mail or a cell phone text message. It is helpful to incorporate financial management software into demand scheduling to track billing, payments, and available account balances.

Automatic demand routing schemes, also known as *feedforward control*, have been developed through continuing research. These schemes are seeing increased use through software that utilizes order information from demand management software to compute a schedule of flow changes to route deliveries through the system. These flow changes are communicated to remote sites through the SCADA infrastructure.

Another possible component of the base unit software suite implements *centralized automatic feedback control*. Similar to the feedforward component, software of this type is seeing increased use. As mentioned earlier, some HMI packages offer internal functionality for implementing varying levels of centralized control. Depending on the type of control, it may be appropriate to run the control functionality as a separate program as the HMI infrastructure may not easily accommodate the required program structure. In this case the external application interacts with the HMI to accumulate necessary measurements and to implement the calculated control actions.

Additional Base Unit Infrastructure

Security Both the physical and IT security of vital infrastructure have come to the forefront in recent years. The appropriate level of each type of security varies from one situation to the next. In some instances it might be advisable to isolate SCADA IT infrastructure on a separate sub-network.

Power Maintaining power to the base unit is imperative. Many irrigation district offices are located in remote locations that are prone to intermittent power outages, whereas others are centered in or near large population areas with seemingly stable power. In either case, it is advisable to have all base unit equipment attached to some sort of uninterruptible power supply. This might also include an emergency backup generator.

3.3. CONTROL OPTIONS IN SCADA SYSTEMS

Supervisory control is the historical foundation of SCADA and is omnipresent in all modes of SCADA operation. One or more supervisory operators remotely manage the system from the operator console(s). Even in a fully automated system, supervisory operators typically are able to intervene or make changes.

Automatic control options can be categorized into several basic groups. Local control is functionality that could be implemented in a stand-alone fashion within the remote unit hardware. Central control is processed at the base, typically using information from multiple remote sites, with control actions being sent to remote units for implementation. Distributed control is implemented at remote sites that possess the ability to communicate with each other. Automatic control concepts are detailed in Chapter 6.

3.3.1. Local Control

Conceptually, local control can be thought of as an autonomous control routine that relies solely on local information immediately available to a PLC at a given remote unit. The control can be enabled and disabled by base unit operators or base unit control software. Local control parameters can be modified by the base unit as well. Field operators can manage local control through the OIT. Examples of local automatic controls include local flow control and local upstream level control.

Many HMI vendors provide the structure to implement local control loops on a timed basis within the software at the base unit. The manner in which these controls are implemented varies from one HMI vendor to the next. This implementation can be thought of as local control implemented centrally. This arrangement can be advantageous when first implementing automatic control in a system. First, it can be helpful for diagnosing issues in the control design. Second, it allows supervisory operators to gain experience in the response of the local control routines to various scenarios before the control is deployed locally in the PLC. A disadvantage of this centralized implementation is an increased load on the communications infrastructure. Additionally, a reliable communications infrastructure is imperative.

3.3.2. Central Control

In centralized control, required data items are collected from multiple remote locations by the base unit. Control actions are calculated centrally, and the resulting control actions are sent to individual remote sites for implementation. This paradigm allows the implementation of robust, globally tuned central control configurations allowing a canal to be treated as a complete system instead of a series of individually controlled pools.

Peer-to-peer architectures can allow for the implementation of some central control schemes at the local level. Although this arrangement can be advantageous from communication and infrastructure perspectives, care must be taken to ensure proper oversight at the base unit.

3.3.3. Distributed Control

Peer-to-peer architectures can allow for the implementation of some multi-site control schemes at the local level. This distributed control arrangement can be advantageous from communication and infrastructure perspectives. However, care must be taken to ensure proper oversight at the base unit.

3.4. SCADA PROJECT CONSIDERATIONS

3.4.1. System Evaluation

A reasonable first step toward implementing a SCADA system is a thorough evaluation of the overall operation of the canal system. The goals of this evaluation are to determine what operational changes are necessary or desirable in the system, whether SCADA is the best approach to implementing these changes, and what specific functions the SCADA system could accomplish. Stakeholder input is crucial in this evaluation. Operational changes may require modification to ordering procedures. Customers may need to operate under a different set of rules that could affect their order scheduling.

3.4.2. Building an Integration Team

When implementing a SCADA project, an organization has a number of options ranging from utilizing in-house personnel to purchasing a complete system from a commercial vendor. Doing a project in-house can be challenging, especially if automatic control is involved or the project is large. In spite of many good hardware and software products available on the market, putting all the pieces together requires specialized experience and skills.

If in-house expertise is not available, building an integration team is an alternative. The required set of skills depends on the scale of the project and the level of automatic control required. A single system integrator may be able to set up a simple supervisory system. A large system utilizing automatic controls may require the skills of an integrator, an automation consultant, civil engineers, communications specialists, and IT professionals.

3.4.3. System Maintenance

Maintenance needs are a consideration when investing in SCADA development. Routine hardware maintenance includes the replacement of failed sensors, old batteries, and compromised wiring. Even with lightning protection, a site experiencing a direct strike may require complete replacement of sensors and the PLC. Vandalism also should be considered when estimating maintenance costs. Although remote site software can remain unchanged for extended periods, occasional upgrades may be prescribed by the manufacturer. Routine software maintenance should be expected at the base. Many times HMI software updates are required to accommodate security updates to the underlying computer operating system. The implementation of automatic control carries its own added level of maintenance to check periodically and tune control parameters.

3.5. SUMMARY

SCADA offers a generous set of building blocks to facilitate the implementation of varying levels of control on a canal system. Whether starting from scratch or building on an existing SCADA platform, the scalable nature of today's hardware and software technologies allows for a phased implementation resulting in a robust management and control system.

Although it might appear simple in a broad view, SCADA development can be a complex undertaking. Whether utilizing in-house expertise or working with contracted experts, a well-thought-out development plan is essential. Establishing standards for equipment placement, wiring, program code, operator interfaces, and HMI database elements can save in development, installation, and maintenance overhead.

Supervisory control may be an initial goal of an implementation, yet varying levels of automatic control are available. Through continuing research efforts, some theoretical control strategies have been tested on actual canals and are now being implemented. Today, it is possible to build a system that is capable of routing flow changes automatically while also maintaining water levels throughout the system. The following chapters document the evolution of this research and present the theoretical information needed to implement an array of automatic control schemes.

During one of the meetings for the production of this manual, Sumith Choy astutely referred to the path through canal automation as a journey. That said, whether your destination is a basic supervisory remote control system or a complex system utilizing full automatic control, SCADA technologies provide the path to get there.

CHAPTER 4
OPERATION AND CONTROL CONCEPTS

4.1. INTRODUCTION

This chapter deals with the different concepts and configurations for operating a network of canals that deliver irrigation water to users. It is important to note that the concepts discussed are not necessarily mutually exclusive because different concepts often can be used on different parts of the same canal network. Each concept has advantages and drawbacks, and each fits better with certain overall strategies.

The main intent of canal networks is to supply irrigation water to farmers to grow crops, although any transportation of water from source to user (e.g., supply to urban areas or transport of drainage water through canals) can be handled with the concepts presented. Experience has shown that farmers can be more productive when they have better control of their water supply, which implies that they are able to

1. Determine when the water will arrive (frequency),
2. Determine the rate of flow to match their on-farm irrigation system (rate), and
3. Turn the water off when they are done (duration).

The overall intent of this manual is to improve the farmers' ability to control their water supply by developing improved canal operations. Methods for control of irrigation canals can be grouped into two categories: *supply-oriented* and *demand-oriented,* although most canal control strategies include aspects of both. For a supply-oriented system, the decisions regarding the amount of water diverted and delivered to users are made by a government agency or central authority. Such decisions define the amount of flow in each successive canal (i.e., main canal, branch canal lateral, sublateral, etc.). Eventually, water arrives to a small group of users who rotate the flow among themselves. Some level of user control can be exerted at that point. Attempts to modernize tend to move the ability to make changes further up the system's management. Providing flexibility to a water user association, say at the lateral canal level, is often part of modernization efforts.

Demand-oriented systems allow farmers to order water when required for their farm management. No system allows farmers to take an unlimited amount of water at any time. Canal and outlet capacities limit flows, and in some cases the water supply itself may be even more limiting than the canal capacities. A few limited rate demand systems exist, such as where reservoirs are available to provide storage, but most flexible irrigation water distribution systems are arranged systems, where the farmer must order water from the water purveyor.

Often water supplies are limited, and each user has a limited share of the available water. Thus, even though these systems are demand-oriented, they must operate within the constraints of the available water supply, both for capacity and water rights. Even for demand-oriented systems, operators tend to limit large demand variations by moving water orders a day earlier or later. Large, frequent fluctuations in flow cause the system to be in a transient state for longer periods, which can cause flows to fluctuate. Fluctuating flows generally are not desirable from the farmers' perspective because they complicate on-farm water management. It is not possible or reasonable to adjust the supply in a canal, because often there is a significant time lag between release of water at the source (e.g., reservoir) and delivery of water to users. This method also encourages operators to maintain more uniform flows at the upstream end of canal systems.

These constraints on water supply can have a significant influence on the canal control strategy. Whether the water is scheduled by a central authority or based on farmer water demands, the rate of water release from the water source is sometimes a decision that is independent from pure demand considerations. Further, flow rate measurements always have some error, such that even when the total measured supply rate matches the actual user demand, it is still challenging to match supply and demand for each canal down the canal network, as errors can occur at each branch point. The net result is that most individual canals almost always have a situation where supply and demand do not match. The canal control strategy essentially provides the strategy for dealing with these supply-and-demand mismatches.

4.2. GENERAL STRATEGIES

The assumption during the traditional design of canal networks for water supply is that the operator is able to set the correct flow at each bifurcation (i.e., an offtake location) to meet the operational requirements. If the operator were able to do that, no further control would be necessary. This is one reason that measurement and control of flow rate are so fundamental to canal control. When flow changes are made in a main canal, the operator needs to adjust each outlet and check structure as this flow increase passes. Because flow changes arrive gradually downstream, this already becomes a difficult task, because the operator cannot always be at each gate and adjust the gate positions as the flow increases. A common approach is to make gate adjustments at each site successively and then return to each gate to make corrections, depending on the observed situation. These initial changes and the later corrections can be interpreted as *feedforward* and *feedback* control actions, respectively. The initial changes in settings to provide the new flows downstream are feedforward actions, whereas the corrections based on the actual situation are feedback actions. Remote control, where observations are communicated to a central control room and, from there, gate adjustments can be remotely implemented, simply allows the users to perform these actions more quickly, more often, with better timing, and with the ability to see more of the canal conditions at once. Automatic controls allow these adjustments to be made automatically. Automation allows control actions to be made more often and, in theory, more precisely. But fundamentally, these different approaches (manual/local, manual/remote, automatic) use the same basic strategies.

For a given canal, the control strategy is simply how to deal with the mismatch of supply and demand, including those created internally by poor measurement and control of flow to users. In reality, there are only four options:

1. Fixed structures at the outlet force the errors in flow to be absorbed by the outlets. At the lower end of a canal system, this is the delivery to the farmer. This is the most common approach but results in poor utilization of supply and so-called tail-ender problems.
2. Allow the mismatch to go downstream, where the spilled water can either be recaptured or lost to the system. This strategy generally uses upstream control, where the water level upstream from each structure helps to provide the correct flow at each outlet. As flow shortages are common, the inflow to these systems generally exceeds the demand to ensure that the last user receives enough water, which results in spills.
3. Move the mismatch upstream, where it can be stored or absorbed into the system. These strategies generally use downstream control, where the control observes water levels at structures downstream and changes the inflow to balance supply and demand. In this strategy, the canal inflows are adjusted to better match demands. This strategy assumes that the excess flow can be absorbed upstream, for

example, in a reservoir. Delivery through pressurized pipelines functions in this way. When the downstream tap is opened, the pressure in the pipeline delivers the water and the upstream pressure in the reservoir keeps the pipeline filled, given enough pressure and water availability.

4. Provide storage within or adjacent to the canal where mismatches can be stored temporarily. If canals are large enough, they can provide significant storage (that is, function as reservoirs). With regulating reservoirs adjacent to the canal, this strategy then could use upstream control in the canal upstream from this reservoir and downstream control in the canal downstream from this reservoir.

4.3. CONTROL DIAGRAMS

In this section, each basic control concept is explained using a control block diagram and a schematic of a single canal pool. These basic concepts are applied to segments of a canal network. They have to be organized and applied to different segments of the network to provide an overall control scheme. Following the control diagrams, common configurations of control strategies are discussed, along with advantages and disadvantages (Malaterre et al. 1998).

The general control diagram (Fig. 4-1) provides the components of an individual control action. Each block in the diagram represents one component of the control process. On the left-hand side is the target value or set point for the control. This control variable can be a water level, flow rate, pressure, volume, or others. This is compared to the actual value of this variable (on the far right side), and the error in that value is determined (actual minus set point). A control decision is made based on the error, and a control action is made to the structure (gate, actuator, or pump). The control decision is to move the structure a certain amount, and the control action physically makes that change. The change in the structure causes the flow in the system to change, which influences the conditions in the canal pool. External disturbances also can alter the conditions in the canal pool such as an unexpected change in inflow to the canal pool or a change in the outlet flow. The combined influences of the control action and the disturbance are called the *output*, which are simply the conditions within the pool. The control variable is observed, and the observed value is the input to the control action. The control variable and control action are different for different types of control, but these basic features are included in any control system, whether manual or automatic. Such diagrams help to define the conditions and requirements for each control system.

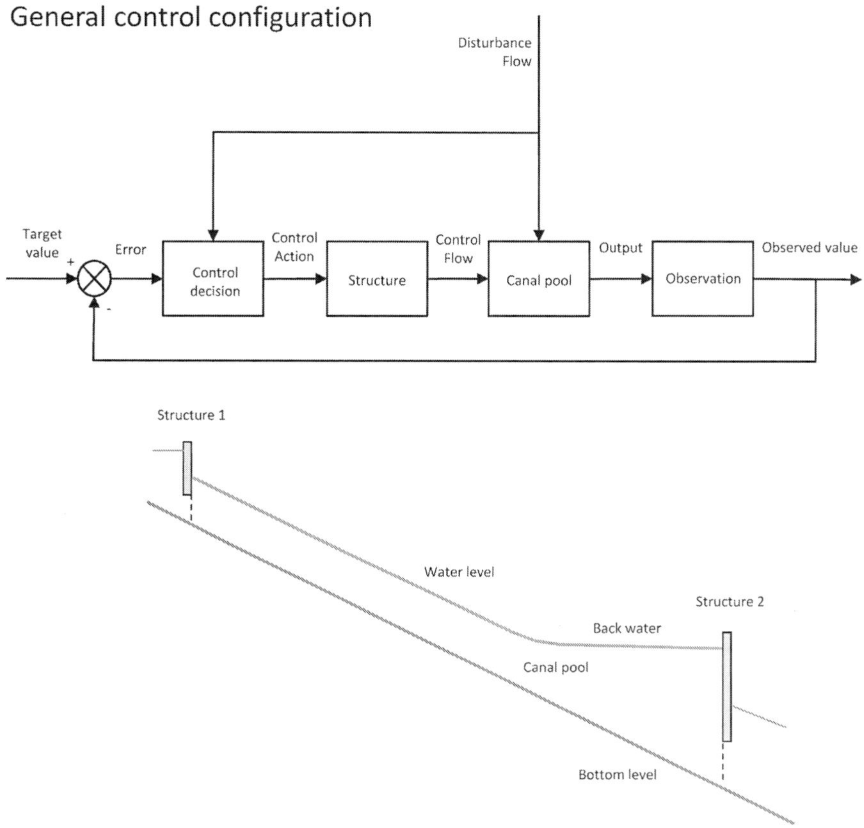

Fig. 4-1. General control configuration diagram

4.3.1. Manual Control and Automatic Control

Canals or parts of canals can be controlled by operators or by computers. In the first case, a local operator does a visual reading of the water level that needs to be controlled and, if necessary, takes action by changing the setting of the structure. This control concept (Fig. 4-2) is commonly referred to as *manual* control. Manual control also can be implemented as remote manual control, using a properly designed SCADA system.

In the case of *automatic* control (Fig. 4-3), electronic sensors measure the water variable and send a signal to a microprocessor or programmable logic controller (PLC). An analog-to-digital convertor makes the signal readable for the computer. An algorithm calculates the required control action and sends this through a digital-to-analog convertor or a digital on-off signal to the motor of the structure. The motor actuates the required change. Usually, the automatic loop is repeated at a fixed time interval.

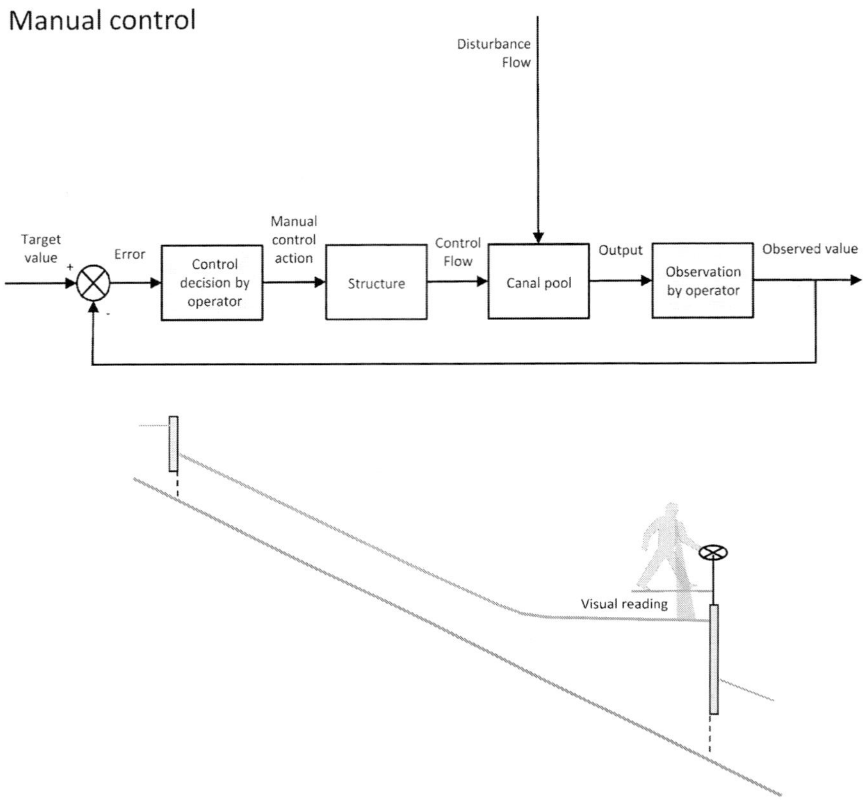

Fig. 4-2. Manual control configuration diagram

In Fig. 4-3 the automatic control is depicted as *local control*. Another non-manual way of controlling the gates is by hydraulic (float-actuated) gates such as AVIS, AVIO, or AMIL gates. These types of hydromechanical gates are discussed briefly in Chapter 2 but not further considered in this manual.

4.3.2. Feedback Control and Feedforward Control

Feedback control is a control methodology that compares an observed water variable value with a target value. When there is a deviation between the observed and desired value, a controller calculates a required change for the input of the canal pool, based on this error. The required change is sent to the structure that implements the required change, resulting in an adjusted structure flow that influences the hydraulic conditions of the canal pool. The control loop is closed, because the controller

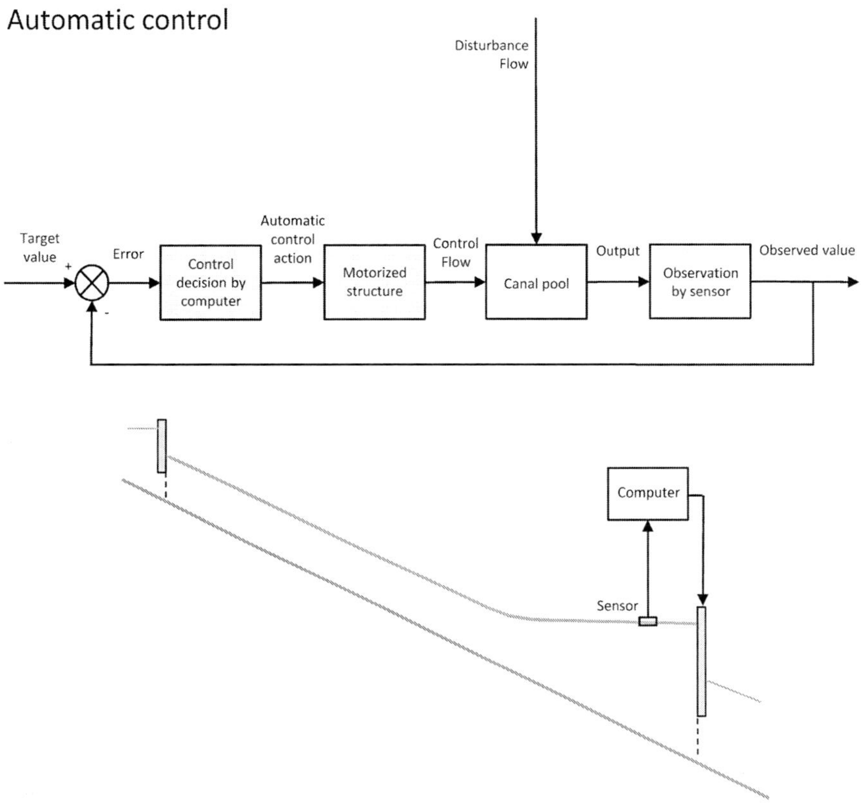

Fig. 4-3. Automatic control configuration diagram

observes the resulting change in water level in a future time step, and the control loop is executed. In Fig. 4-4 distant downstream feedback control is presented.

Unlike feedback control, *feedforward control* does not react to an error between the target and observed water level but to a disturbance that could create a future error (e.g., a future water delivery). To counteract the effect the disturbance has on the water level, the feedforward controller uses a model of the response from which control actions are determined that minimize the effects of this disturbance. In Fig. 4-5, this model is merely the estimated delay time (represented as τ_d) between Structure 1 and the turnout location. When it is known that the turnout structure will open at time t with a flow of Q_d, Structure 1 can start releasing this extra amount Q_d at time t-τ_d (anticipated). Feedforward controllers often use anticipation of the ordered offtake flows (see Chapter 6).

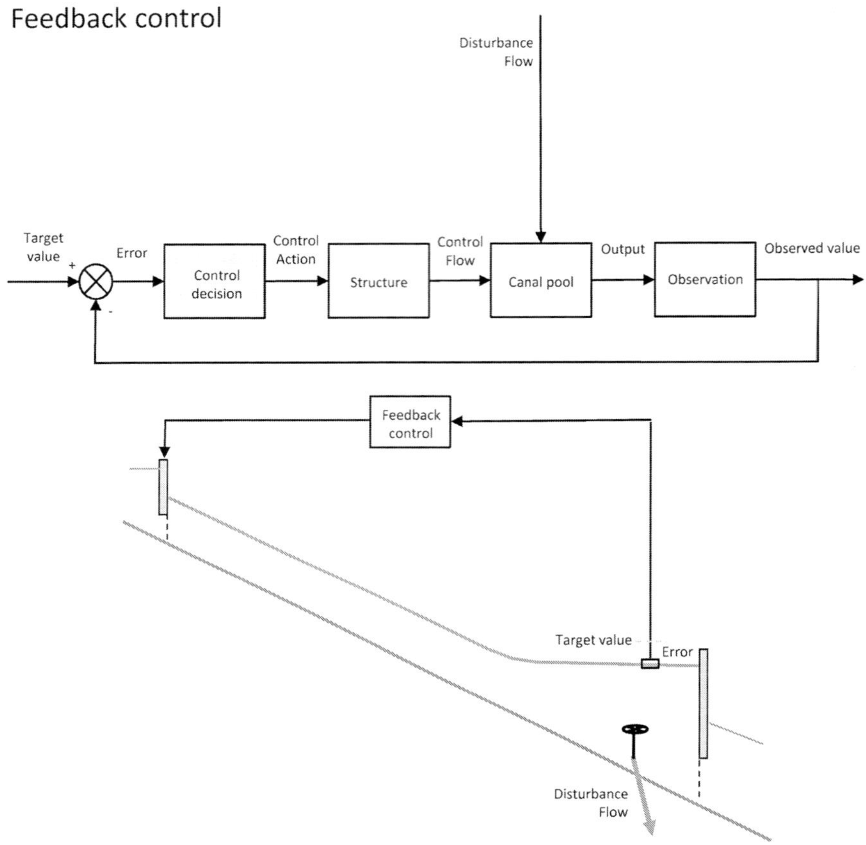

Fig. 4-4. Feedback control configuration diagram

4.3.3. Upstream Control and Downstream Control

When using feedback control of canal water levels, two types of control directions can be used: *upstream* or *downstream*. The terms *upstream water level control* and *downstream water level control* are shortened to upstream control and downstream control. When using upstream control (Fig. 4-6), the observation is made upstream from the control structure. When an increase in flow arrives from upstream, a change in water level is detected and the controller passes this disturbance (change in flow) to the downstream pool.

With downstream control (Fig. 4-7), the observation is located downstream from the control structure. A change in water level there, for example a water level drop, is detected and results in a change in the

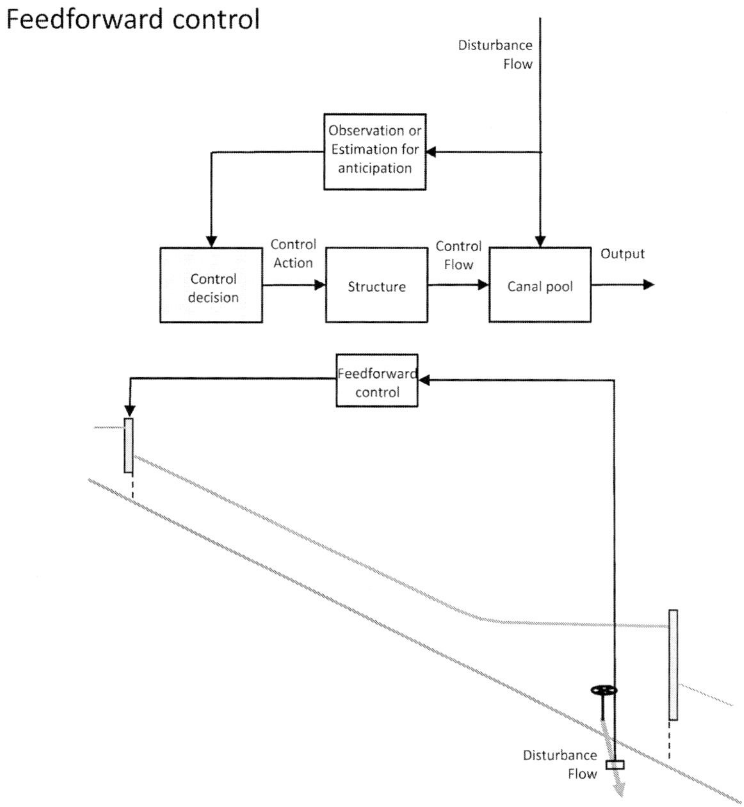

Fig. 4-5. Feedforward control configuration diagram

opening of the upstream control structure. In the example, the change is an increase in flow to bring the water level back to the water level target.

4.3.4. Local-Observation Control, Distant-Observation Control and Remote Control

Feedback control can be applied using an observation at any location in the canal reach. Local-observation control (*local control*) (Fig. 4-8) refers to the concept where the observation is made in close proximity to the control structure.

In cases of distant-observation control (*distant control*) (Fig. 4-9), the observation is located a significant distance away from the control structure, usually at the other end of the canal pool. Unlike local control, communications are necessary to transfer the observation signal to the other end of the pool.

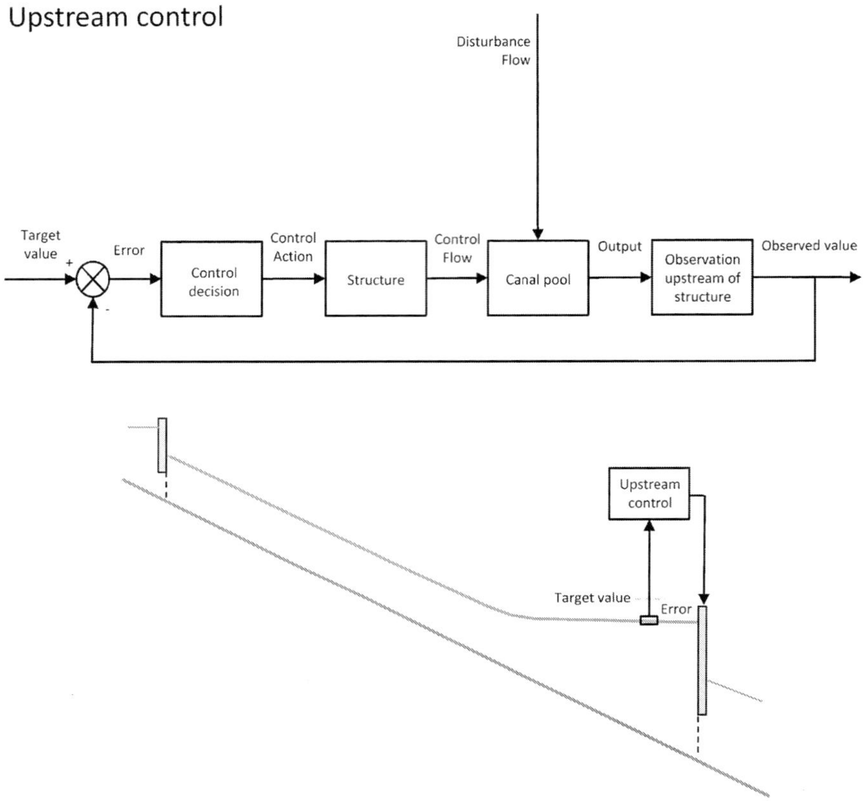

Fig. 4-6. Upstream control configuration diagram

Finally, *remote control* (Fig. 4-10) is the concept whereby all actions are decided and sent from a central control center.

4.3.5. Control of Water Level, Flow, or Volume

Various water variables can be controlled. These are referred to as *controlled variables* and may include water level, flow rate, and volume. Water level control is the most common and was discussed previously. Figure 4-11 shows the difference between water level control and *flow control*, in this case, feedback flow control.

The control of flow can be executed by using feedback control as depicted in Fig. 4-11 or by feedforward control, where the head-discharge equation of the structure is inverted to put the structure setting in a position that should correspond to the required flow. In the example given in Fig. 4-12, the water levels that influence the structure flow are observed,

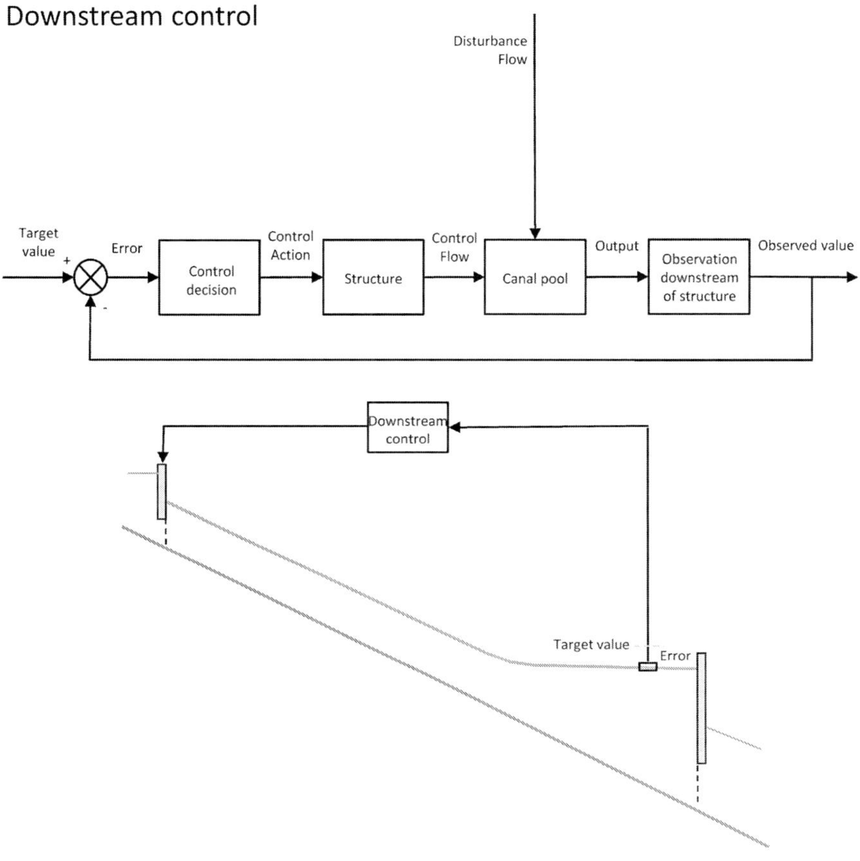

Fig. 4-7. Downstream control configuration diagram

and the gate formula is used to calculate the proper setting of the gate position. The control action variable in the first case is flow, whereas in the second the control action variable is the gate position. Flow control is applied to the flow through a structure, thus, to the canal downstream. The flow measurement device is often downstream of the structure. Often it is very near to the structure but in some cases could be very far (e.g., a spill at the downstream end of the canal). Upstream control of flow rate is simply not a relevant concept.

As volume cannot be observed directly, *volume control* (Fig. 4-13) estimates the volume based on water level observations, usually one at the upstream end and one at the downstream end of the canal pool. The volume can be estimated using the dimensions of the canal pool and a weighted average of the water level observations, or from backwater

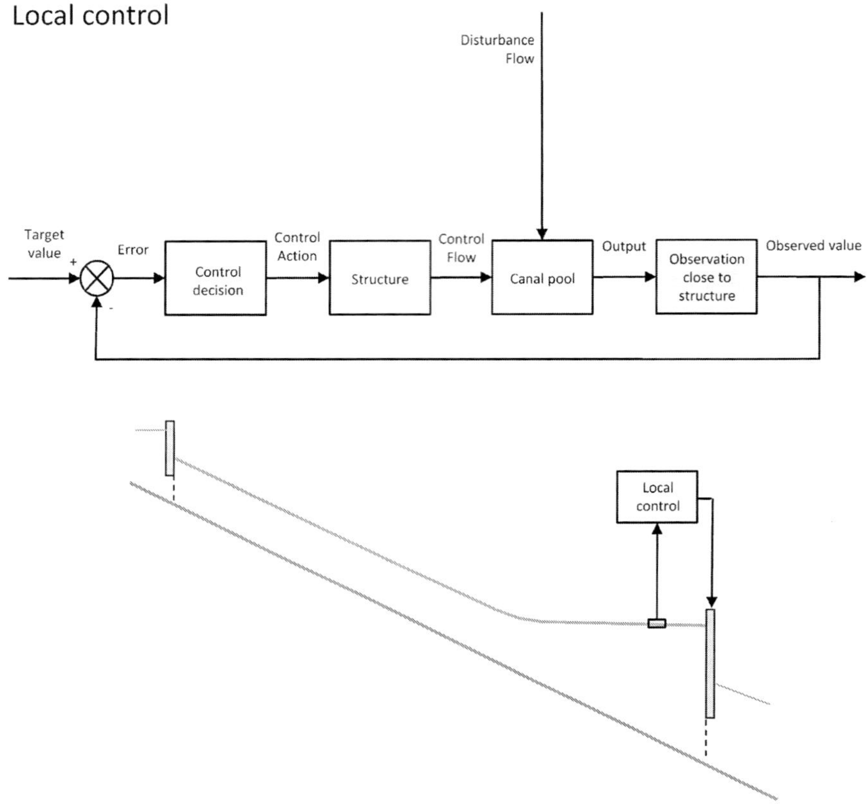

Fig. 4-8. Local control configuration diagram

calculations. Constant volume control maintains constant canal pool volumes. Because of changes in the hydraulic grade line, the water levels at most locations in the canal change with flow rate. Near constant levels are provided somewhere near the center of the pool. Because most turnouts are not near the center of the pool, turnout gates must be adjusted to maintain a constant discharge, typically with constant flow controllers. Volume controllers that allow pool volumes to change over time allow the canal to be used for temporary storage of water. The different concepts are summarized in Table 4-1.

4.4. COMBINATION OF CONTROL CONCEPTS

For a network of canals, the control concepts as discussed are used in combination. To illustrate this, traditional control is detailed here. Under

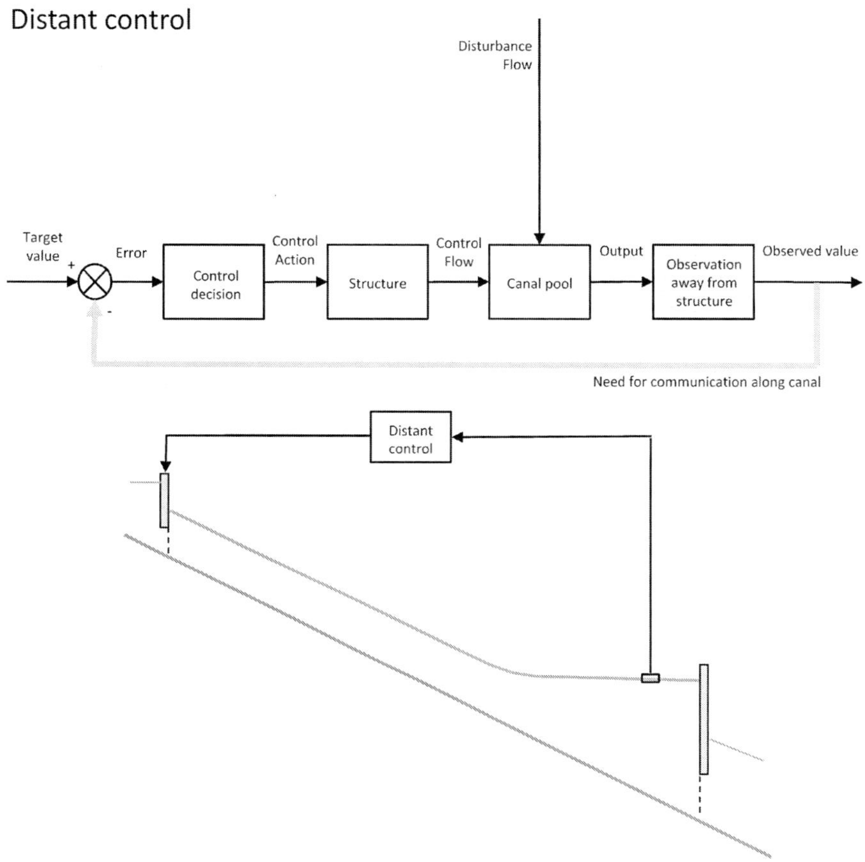

Fig. 4-9. Distant control configuration diagram

manual control, the flow rate entering the canal is set by an operator, whether it matches the actual demand downstream or not. The operator makes gate adjustments (both offtake and continuing canal structures) at each bifurcation to divide the water properly. With traditional manual control, these gates remain fixed, and flow is divided according to the hydraulics of the structures (i.e., the relative sensitivity of gate flow to a water level change). After some time (e.g., once the water levels have stabilized), the operator returns to the canal to determine if these control settings are distributing the water properly. If not, the operator makes additional adjustments. The canal inflow and initial gate changes are feedforward control, while future adjustments are local upstream feedback control. Flow control can be applied to the canal headgate. Automatic gates or mechanical structures that provide constant (or near

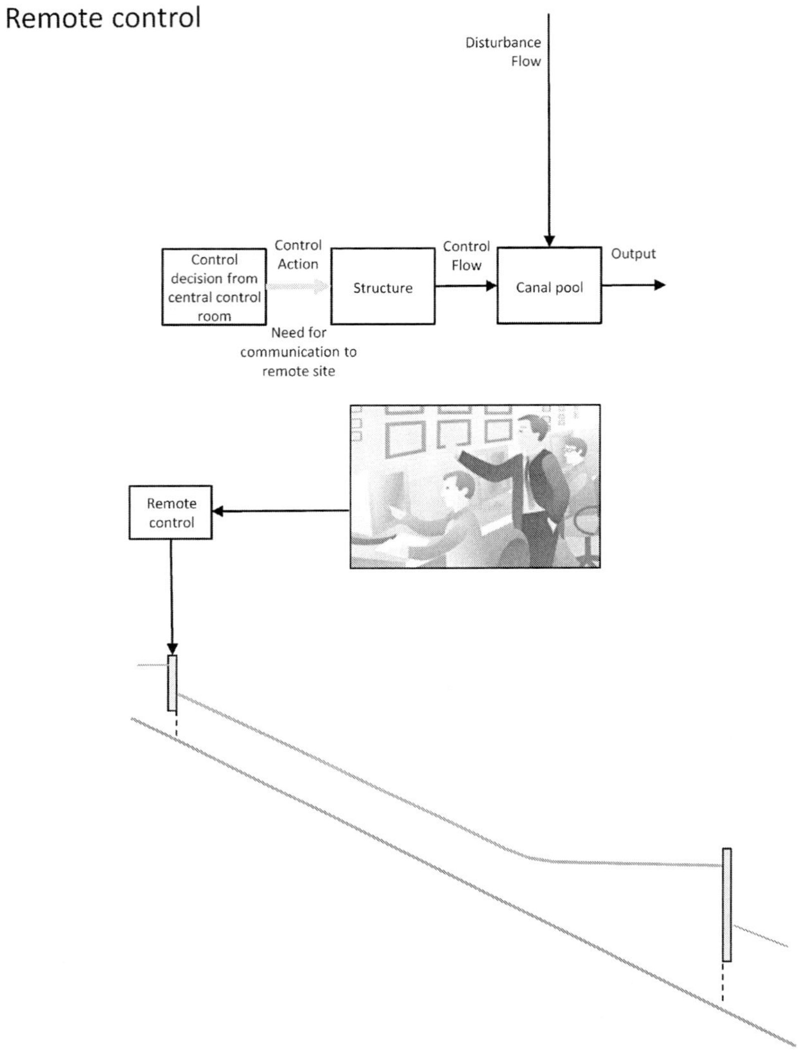

Fig. 4-10. Remote control configuration diagram

constant) upstream levels provide more constant flow to the canal turnouts, even when those turnout structures are not automated. Any deficit in canal inflow results in a shortage at the last turnout. A surplus either goes to the last turnout or spills. If the operator is allowed to change the canal inflow to increase or decrease the canal inflow, when too high or low, this can be considered a form of downstream control (e.g., if water levels or flows are in error at the downstream end of the canal).

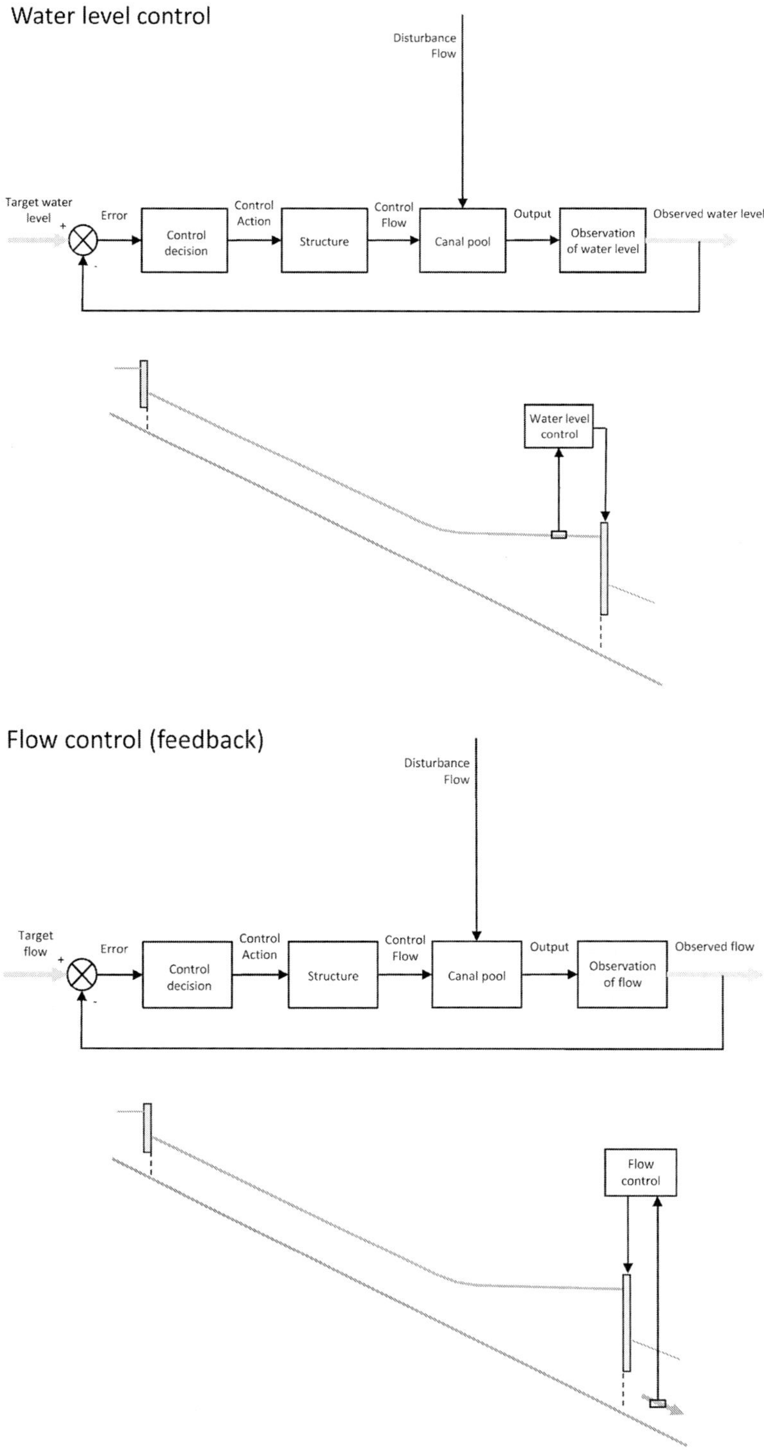

Fig. 4-11. Water level and flow control (feedback) configuration diagrams

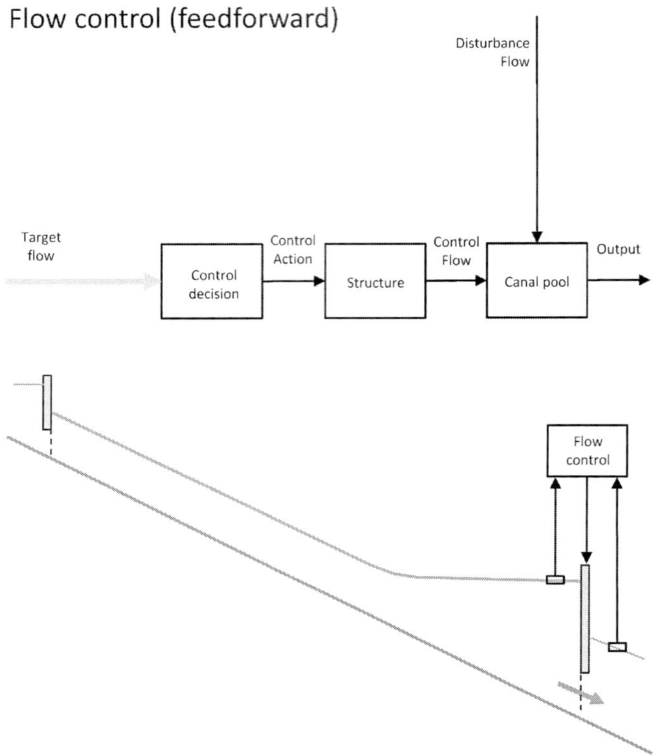

Fig. 4-12. Flow control (feedforward) configuration diagram

When going from traditional control to canal automation control, these control concepts are combined as well. Here are some examples:

1. Flow control at the canal headgate, upstream control at all check gates downstream, and (with or without) routing of demand changes through the structures;
2. Downstream (local) control of the canal headgate and all (except the last) check structures downstream; note that level-top canals are required for this control method;
3. Downstream (distant) control of the canal headgate and all (except the last) check structures downstream and routing of flow changes through the structures; and
4. Volume control of the main canal check structures, in combination with numbers 1 or 2 for lateral canals.

There are many more options to fit the requirements of each district.

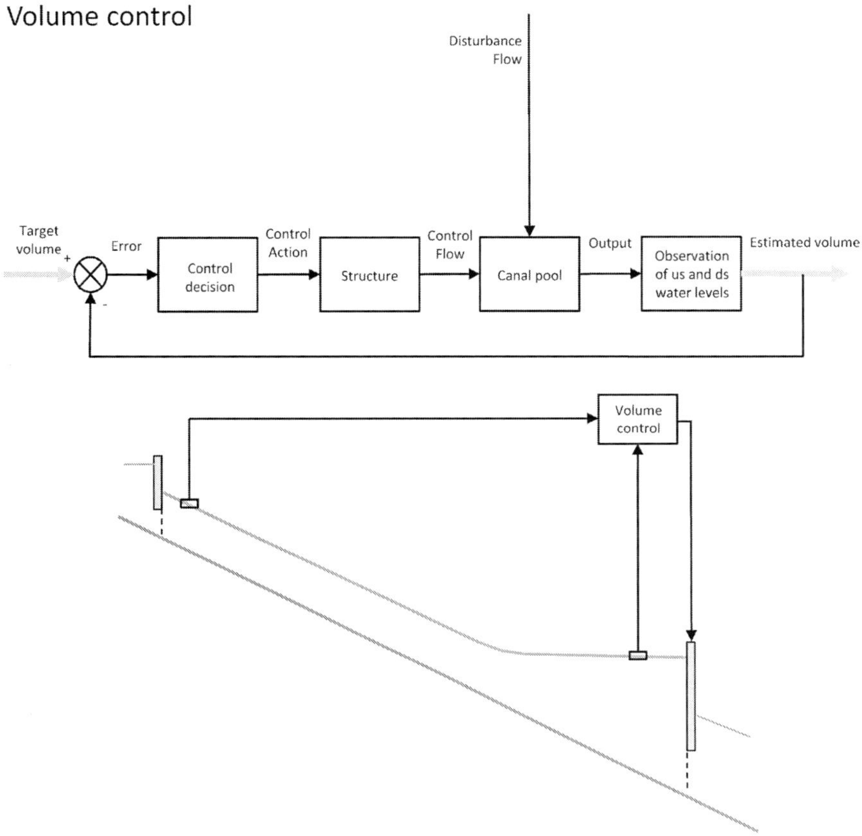

Fig. 4-13. Volume control configuration diagram (us and ds: upstream and downstream)

Table 4-1. Summary of Different Concepts of Control

Mode of operation	Manual	Automatic
Control method	Feedback	Feedforward
Location of action	Local	Remote
Proximity of measurement	Local	Distant
Direction of control	Upstream	Downstream
Control action variable	Gate position	Flow
Controlled variable	Water level and volume	Flow

4.5. NETWORK CONTROL WITH CONTROL SEGMENTS

Often it is useful to set up different canals or canal reaches as separate control segments. These segments are simply a group of canal structures that control one part of the network. Typically, the main canal would be one canal segment, and the lateral canals can each be separate control segments. For example, the main canal can be operated with manual remote control, and the lateral canals could be operated with automatic upstream level control. The upstream end of the lateral canal typically would use flow control. Another obvious choice for the division into different control segments is the location of a reservoir. The canal reaches upstream from the reservoir can be in upstream control, whereas those downstream from the reservoir can be set to downstream control.

The control of flow or water levels within one segment may be influenced by controls within another segment. Thus the control or transfer of information between segments must be identified clearly.

4.6. REFERENCES

Malaterre, P. O., Rogers, D. C., and Schuurmans, J. (1998). "Classification of canal control algorithms." *J. Irrig. Drain. Eng.*, 115, 701–713.

CHAPTER 5
CANAL HYDRAULIC PROPERTIES

5.1. INTRODUCTION

This chapter discusses the hydraulic properties of canal pools and structure as they relate to needs for canal automation. As this manual of practice is geared toward modernization of existing canal networks or systems, detailed design considerations are not included. However, changes may be needed to the infrastructure to implement new technologies during modernization so that the system will allow for greater flexibility and, thus, potentially improve performance.

Sometimes, modernization results in the replacement of canals with pipelines. Although a modernized system may have pipelines and canals, control of pressurized pipelines is outside the scope of this manual. Low-head or gravity pipelines often use the same control structures as canals. Canals generally are much more difficult to control because of time delays and wave dispersion, as discussed in this chapter. Gravity pipelines generally have shorter time delays and little or no wave dispersion and, thus, are simpler to deal with from a control perspective.

Water conveyance and delivery systems vary greatly in size and complexity and may include some or all of the following components (see Glossary for definitions):

- Storage reservoir;
- Regulating or buffer reservoir;
- Diversion dam;
- Distribution system;
- Canal check gate structure;
- Canal pool;

- Turnout, outlet, or offtake;
- Pumping plants;
- Inverted siphon; and
- Wasteways.

The main focus of this chapter is the influence of canal and structure hydraulics on canal operations. This includes understanding the hydraulic properties of canal check structures, how flow changes propagate through a canal, and how canal water levels and volumes change when canal discharge changes. Additional considerations related to canal design may influence what can be accomplished through modernization efforts. To provide additional background, design issues are discussed briefly first in Section 5.2.

5.2. DESIGN ISSUES

Open-channel systems carry water under the action of gravity from the source to the delivery point. Pumps may be used to lift water between canal pools. Check structures usually are used to provide greater upstream water surface elevations than would otherwise be the case. The extent of operational improvements possible in a system can be limited by the water flow rate that can pass through the canal. The flow rate that can pass through the canal is limited by the canal cross section, longitudinal slope, and hydraulic resistance, as well as by hydraulic structures (check structures, siphons, etc.) and freeboard.

5.2.1. Freeboard

By design, canal lining and canal banks usually are extended above the maximum water level as a safety measure to prevent the canal from overtopping due to operator error or unforeseen conditions. This extended distance is referred to as *freeboard* and often is required for three main reasons:

1. An uncertain flow rate entering the canal; additional depth may be required to accommodate a flow rate that is above the design value;
2. Uncertain canal hydraulic roughness; this is the most common reason that freeboard is required (because algae, weeds, sediment, deteriorating canal lining, and deposits on canal walls can all create higher hydraulic losses and result in a greater water depth for a given discharge); and
3. To accommodate the passage of waves safely during changing flow conditions.

Modernization and the associated improvements in maintenance practices can reduce the required freeboard. Canal operators quickly learn what flows can be conveyed safely by a canal and how these flow limits may change during the season (i.e., in between canal cleaning or maintenance).

For small canals, the U.S. Natural Resources Conservation Service (formerly the Soil Conservation Service) recommends a freeboard of 20% of the design flow depth (USDA SCS 1977). For larger canals, the U.S. Bureau of Reclamation has developed guidelines in graphical form (see Fig 1.9 in Aisenbrey et al. 1978). For unlined canals, the freeboard determines the height of the earthen lining. For lined canals, freeboard determines the height of the lining, but the canal bank also is extended above the lining to provide additional protection.

5.2.2. Canal Lining

Canals often are lined under the assumption that lining will reduce water losses. For some canals, the seepage is insignificant, and lining the canal will not save a significant amount of water. However, lining usually changes the canal prism and results in a smaller cross section. Water often travels faster in a lined canal, thus making the canal potentially more responsive to downstream needs. Also, lined canals tend to require less maintenance, in general. Lack of canal maintenance is a common problem in canal systems that have not been modernized. For canals that have significant seepage, compaction of the soil at the bottom and sides of the canal prism can be a cost-effective alternative to lining.

Common reasons for canal lining include

1. Saving water (reduce seepage);
2. Stabilizing channel bed and banks (reduce erosion);
3. Avoiding piping through and under channel banks;
4. Decreasing hydraulic roughness (flow resistance);
5. Promoting movement, rather than deposition, of sediments;
6. Avoiding water-logging of adjacent land;
7. Controlling weed growth;
8. Decreasing maintenance costs and facilitating cleaning; and
9. Reducing excavation costs (when extant material is unsuitable).

5.2.3. Intermediate Structures

Check structures force the water level in the canal to be a certain distance above the land surface or above the elevation of lower level canals so that water can be delivered effectively through offtake structures. Other structures in the canal also can influence the amount of water that can be conveyed and the speed at which canal flows and water levels respond to flow changes. Canals often cross streams, rivers, roads, railroads, and other geographical or developed structures. If the flow passes over these features, for example through a flume, the cross section of the flume can limit flows that can be conveyed. Other contractions in the flow area, such as an intermediate weir, also can limit flows. If the canal passes under the feature, such as through a culvert or inverted siphon, the size of the siphon can influence the canal capacity significantly, as well as influence the responsiveness of the canal. These intermediate structures have to be considered explicitly when developing a model of the canal's response to disturbances or flow changes.

5.2.4. Flow Capacity

When an irrigation canal is designed, the conveyance capacity generally is based on the maximum crop water demand for the area serviced by the canal. In cases where the water supply (volume or available flow rate) is not sufficient to meet the consumptive use of crops, then potential crop yields or quality may be compromised. When calculating the crop water delivery requirement to design the conveyance capacity of a canal that supplies water from an irrigation canal to farms (or group of farms), all pertaining factors must be considered, including the daily water use of the crop, area irrigated, application efficiency of the farm irrigation system, and delivery discharge.

The modernization methods described in other chapters of this manual put the spotlight on improving crop production by allowing more *demand-oriented* water deliveries, which potentially can reduce periods of water stress but require larger canal capacities so that farmers can be more flexible in their ability to receive water. In some cases, the water flow rate is determined so that it exactly meets the maximum farm water demand. Considering the frequency, rate, and duration of water deliveries at all levels and the location of the decision making, Clemmens (1987) classified three common delivery methods, which can influence the required capacity of the conveyance system heavily:

- *Centralized scheduling.* The users do not order water. The central authority determines when and how much water is to be delivered. The characteristic method of this group is some form of rotation. The most rigid rotational delivery schedules provide water with no flexibility in frequency, rate, or duration. In the simplest form, a fixed flow rate arrives at a certain date and time on a calendar basis and for a fixed duration. The resulting equation, Eq. (5-1), for a rotation system based on continuous flow at the offtake or turnout during the delivery period is

$$Q_t = W_u A_t \tag{5-1}$$

where Q_t is the flow rate delivered to the turnout, W_u is the gross water use rate required by the crop (including farm irrigation losses), and A_t is the area served by the turnout.

Some rotation systems use the maximum monthly crop consumptive use based on the "average" crop mix. This method can result in water stress when the crop mix differs substantially from the average or when there are periods (e.g., weekly) of higher than usual crop water consumption. Thus, some care must be taken in how W_u is estimated. In practice, there are many variations of rotational delivery schedules.

- *Arranged schedule.* Water is requested by the user for specific date and time, as well as a specific flow and duration and is arranged between the central authority and the user. When possible, the central authority accommodates the request.
- *Limited rate demand.* This is the delivery method generally provided to urban homeowners for a domestic water supply. There is no need to make any request for water. It is available at all times. The maximum flow rate is limited, for example by the size of the pipe branch servicing individual homes or by pressure. This kind of service also can be made available for agricultural irrigation if the flow rate is restricted and water is taken from a location where adequate storage is available. It provides a higher level of water delivery service to farmers, compared to the other discussed approaches, but typically at a smaller rate of flow.

The conveyance capacity of an irrigation canal differs according to the method of water delivery. For convenience, Q_n is defined as the capacity of the canal relative to that required for a rotation schedule, or $Q_n = Q / Q_t$. Similarly, we define A_n as the ratio of the area that could be serviced by flow Q_t based on rotation, or $A_n = A / A_t$. For a rotation system, the resulting capacity requirement is thus Eq. (5-2):

$$Q_n = A_n \tag{5-2}$$

For more flexible delivery schedules, the relative capacity is greater than that required for rotation, or $Q_n > A_n$, as discussed following.

Clement's Demand Formula Clement (1965) developed equations for determining demand based on a binomial distribution, where water flow was either on or off, with F as the probability that flow would be "on" during a given day. The formula is based on the probability P that users will be able to be served by the canal at the given capacity. The equation can be written in the current notation as Eq. (5-3):

$$Q_n = A_n + U\sqrt{A_n(1-F)} \tag{5-3}$$

where U is the number of the standard deviation from a normal distribution for a probability P (i.e., at $P = 0.99$, $U = 2.33$; at $P = 0.975$, $U = 1.96$; at $P = 0.9$, $U = 1.28$, etc.).

Equation (5-3) is intended for a system with less than 100 delivery points. An alternative formula is provided for a larger number of delivery points (Clement 1965; Clemmens 1986).

Clement's formula is intended for demand based on a binomial distribution, with a large number of small offtake flows. As such, it is more applicable to sprinkler irrigation and frequent irrigation events. For longer duration events that occur less frequently, demand does not follow a normal distribution. The extreme case is surface irrigation where irrigation might occur once in two weeks. For the days in between, there is no chance of irrigation. So Clement's formula does not consider the delay between irrigation events. This can have a major influence on capacity.

Clemmens's Demand Formula Clemmens (1986) developed expressions for the application of Clement's formula through a Monte Carlo simulation of surface irrigation demands based on a mix of crops, soils, and weather conditions. This resulted in demand formulas that are more useful for irrigation applications, particularly for surface irrigation. The assumption was that irrigation on a particular field would be scheduled when the soil water deficit reaches a certain threshold. For surface irrigation, a 90% probability of having water available on demand was considered adequate. The results were fit to piecewise linear equations:

$$Q_n = 4A_n + 1 \qquad A_n \le 1 \tag{5-4a}$$

$$Q_n = 1.5A_n + 3.5 \quad A_n \ge 1 \tag{5-4b}$$

These capacities are considerably higher than those determined by Clement (1965).

Clemmens (1986) also developed approximate equations for arranged delivery systems:

$$Q_n = 1.6A_n + 1 \quad A_n \le 1 \tag{5-5a}$$

$$Q_n = A_n + 1.6 \quad A_n \ge 1 \tag{5-5b}$$

These equations provide a capacity that generally is below the Clement formula but still greater than the rotation capacities. It is significantly less than the Clemmens demand formula as shown in Eqs. (5-4*a*) and (5-4*b*).

These equations provide for increased canal capacity that allows more water when increased flexibility is required. If on-farm improvements then are made in addition to delivery-system improvements, the reduction in on-farm water demand will provide even more flexibility.

5.3. CANAL STRUCTURE HYDRAULICS

Check structures typically have a combination of weirs and gates that restrict flow so that the water surface elevation remains at a level that allows a water depth sufficient to provide deliveries to an offtake or lower level canal. Generally, the flow approaching a check structure is determined by the upstream structure setting. When the flow arrives at the check structure, the water level rises until the flow through the structures located there equals the upstream inflow. Therefore, for manually adjusted structures, the division of flow depends on structure settings.

If the check inflow and outflow are in equilibrium (that is, inflow equals outflow), a change in inflow will be divided among structures according to their hydraulic properties. This section discusses the properties of various manual structures; the properties of automated structures are discussed in a later section. A common approach for the design of check structures is to include orifice gates that are used to adjust the flow. Offtake structures are also usually orifice gates. Check structures usually include overflow weirs so that excess flows will not cause the canal to overtop. In some cases, these overflow structures discharge into a natural stream, for example, to remove floodwater that enters the canal. In many cases, these overflow weirs pass water to the canal downstream. At bifurcations, overflow weirs generally only pass water to one downstream canal, usually the larger one. Thus, many check structures include both weirs and orifices. Often, however, the crests of fixed overflow weirs are above the normal operating level of the canal.

5.3.1. Weirs

Weir types include sharp-crested weirs, broad-crested weirs, overshot gates, and simple structure headwalls. For free flow, all have discharges that are approximately related to the upstream water level raised to the 3/2 power. Equation (5-6) describes a free-flow relationship between discharge and head for rectangular weirs:

$$Q = \left(C \frac{2}{3} \sqrt{\frac{2}{3} g} \right) b h^{1.5} = C_x b h^{1.5} \tag{5-6}$$

where C is a coefficient; b is the structure width; g is the acceleration of gravity; and h is the water depth upstream from the weir, relative to the weir crest.

The coefficient C includes both the discharge coefficient, which accounts for friction and energy losses, and the influence of the approach velocity. It is close to 1 for rectangular weirs. However, this coefficient is not constant for a given weir but changes gradually and so can be considered constant only over a small range of heads. For simplicity, the constant C_x often is used, but this has units as it includes the gravity term. This equation may not provide sufficient accuracy for flow measurement but can give approximate values for operations. More details on weirs for flow measurement can be found in Chapter 2 of this manual or in Clemmens et al. (2001).

Movable weirs allow simple adjustments for making changes in upstream water level or structure flow. If an upstream water level change of 10 mm is required, the weir is simply raised 10 mm. This change gives approximately the same flow through the structure once the flow stabilizes. If a small change in flow is required, the weir crest is adjusted (i.e., assuming that $C_x b$ is constant) according to Eq. (5-7):

$$h_{\text{new}} = h_{\text{old}}\left(\frac{Q_{\text{new}}}{Q_{\text{old}}}\right)^{2/3} \quad \text{or} \quad \Delta h \approx \frac{2}{3}\frac{h}{Q}\Delta Q \tag{5-7}$$

where the second equation represents linearization of the head-discharge equation relative to the initial head h and discharge Q; Δh is the change in head (in this case weir crest elevation); and ΔQ is the change in discharge required.

Weirs at check gates are seldom submerged. Estimating discharge and the rate of change of discharge with head is less accurate for submerged weirs.

5.3.2. Gates

The discharge through a gate generally is determined by the orifice equation. For free flow through a rectangular orifice, the discharge equation is

$$Q = C\delta wb\sqrt{2g\,(h_1 - \delta w)} \tag{5-8}$$

where C is a coefficient close to unity for rectangular gates; δ is a contraction coefficient; w is the gate opening; b is the orifice width; and h_1 is the upstream water level relative to the gate invert.

The coefficient C includes energy losses and the influence of the approach velocity. It is relatively constant over small changes in head. For free flow through a vertical sluice gate, δ is relatively constant (e.g., 0.61). For radial gates, δ changes with gate lip angle. For submerged gates, the solution is more difficult. For slightly submerged gates, no simple equation is available. For well-submerged gates (e.g., downstream water level relative to gate invert is more than twice the gate opening), the discharge can be approximated by Eq. (5-9):

$$Q = C\delta w b\sqrt{2g(h_1 - h_2)} \quad \text{Fully Submerged} \tag{5-9}$$

where h_2 is the downstream water level relative to the gate invert.

At a small opening, the contraction coefficient in submerged flow is quite close to the contraction coefficient under free flow, but it increases and tends to equalize when the gate becomes largely open.

Submerged gates are more common than submerged weirs because the weir crest is usually significantly higher than the canal invert, whereas gate inverts may be only a small distance above the canal invert. Submergence greatly complicates calibration of gates, particularly when gates operate under free flow during some conditions and under submerged flow in other conditions. The transition between free and submerged flow greatly complicates operations. This transition zone should be avoided if at all possible.

For operations, it is possible to linearize the head-discharge equation of the gate, just as was done for Eq. (5-7). In this case, generally we are interested in the change in gate position required to provide a specified change in discharge. For submerged flow, assuming the head (h_1 – h_2) remains constant in Eq. (5-10)

$$\Delta w = \frac{w}{Q}\Delta Q \quad \text{Submerged} \tag{5-10}$$

where Δw is the change in gate position required to produce a change in flow rate ΔQ, assuming initial gate opening w and discharge Q.

The solution is more complex for the free-flow orifice equation, which gives Eq. (5-11)

$$\Delta Q = \frac{Q}{w}\Delta w - \frac{1}{2}\frac{Q}{(h_1 - \delta w)}\delta\Delta w \quad \text{Free} \tag{5-11}$$

which can be solved for Δw. If the gate position remains fixed, the change in discharge for a given head can be found from Eqs. (5-12*a*) and (5-12*b*):

$$\Delta Q = \frac{1}{2}\frac{Q}{(h_1 - \delta w)}\Delta h_1 \quad \text{Free} \tag{5-12a}$$

$$\Delta Q = \frac{1}{2}\frac{Q}{(h_1 - h_2)}\Delta h \quad \text{Submerged} \tag{5-12b}$$

where Δh in Eq. (5-12*b*) is the change in the relative head, $h_1 - h_2$.

Finally, we can determine the change in head for a given change in gate movement, assuming a constant discharge, with Eqs. (5-13*a*) and (5-13*b*):

$$\Delta h_1 = -2\frac{(h_1 - \delta w)}{w}\Delta w - \delta \Delta w \quad \text{Free} \tag{5-13a}$$

$$\Delta h = \frac{(h_1 - h_2)}{w}\Delta w \quad \text{Submerged} \tag{5-13b}$$

These equations can be used to make incremental corrections in discharge or water level. If larger changes are made, the discharge equations can be solved to provide a better estimate of the change required. In some cases, this may require an iterative solution.

These equations assume steady-state conditions. When a gate position change occurs, celerity waves are generated, which alters the water level immediately upstream and downstream from the gate. These waves influence automatic control devices and so should be considered when water levels or flows are controlled by gate position. These responses are discussed in the section on feedback control. Malaterre and Baume (1997) show how to take account of the instantaneous change caused by gate movement, using the theory of characteristics. This theory gives the link between flow variation and variation of levels h_1 and h_2 in the upstream and downstream pools, respectively, in Eqs. (5-14*a*) and (5-14*b*):

$$\Delta h_1 = \frac{-\Delta Q}{B(c_{d,1} - v_1)} \tag{5-14a}$$

$$\Delta h_2 = \frac{\Delta Q}{B(c_{d,2} + v_2)} \tag{5-14b}$$

where B is the top width; c_d is the celerity of gravity waves (see further, Eq. [5-17]); and v is the mean velocity of flow (subscripts 1 and 2 denote upstream and downstream, respectively).

5.3.3. Examples

Example 1: To determine the gate setting for a desired change in flow, the following data are given: gate width b = 2 m, upstream water level

h_1 = 1.8 m, downstream water level h_2 = 1.0 m, contraction coefficient δ = 0.63, gate opening w = 0.5 m, and C = 1. From Eq. (5-9)

$$Q = 1.0 \times 0.63 \times 0.5 \times 2\sqrt{2 \times 9.81(1.8 - 1.0)} = 2.5 \text{ m}^3/\text{s}$$

To make a change in flow of 0.3 m^3/s, application of Eq. (5-10) gives

$$\Delta w = \frac{0.5}{2.496} \times 0.3 = 0.060 \text{ m}$$

This calculation assumes that the upstream and downstream water levels remain the same, regardless of flow rate. The assumption here is that the inflow from upstream has increased, and the operator is attempting to move the gate to both change the flow rate and keep the water level upstream from the gate essentially the same (for example, at its target level of 1.8 m). The upstream level is usually under backwater from the gate, so the water surface is governed primarily by the check gate opening and the flow coming in from upstream. Unfortunately, the water level downstream is influenced by backwater in the downstream channel. If the downstream channel is heavily under backwater, the downstream depth might not change noticeably because the water surface elevation is governed by the water level at the downstream end of that pool. If the water depth in the downstream channel is governed by channel resistance (that is, governed by friction or normal depth), a change in flow rate will result in a change in water depth.

The initial conditions could result from a downstream channel that is under normal depth and has the following properties: bottom width 3.0 m, side slopes 1.5:1 (horizontal to vertical), bottom slope 0.00032, Manning's n = 0.025. These dimensions give a depth of 1.00 m at 2.496 m^3/s. If the flow rate changes to 2.8 m^3/s, the water depth changes to 1.063 m. This requires a gate opening of 0.584 m, resulting in a change of 0.084 rather than 0.06 m. If the gate is only opened by 0.06 m, only about two-thirds of the desired flow change would occur until the upstream level changed to balance inflow and outflow. The upstream level would have to increase by 0.063 m so that the head difference across the gate is maintained (1.863 – 1.063 = 0.8). Note that the 0.063-m change in water depth and the 0.06-m change in gate opening coincidentally are similar.

If there are other obstructions in the channel, such as culverts, the water depth can change much more significantly and less predictably. Now suppose that the downstream level increases by 15 cm (that is, to 1.15 m) for an increase in flow rate through the gate of 0.3 m^3/s. Solving Eq. (5-9) by trial and error gives a gate opening increase of 0.121 m to achieve the desired change in flow rate while maintaining a constant upstream level. Therefore, the change in the downstream water level caused the operator to move the gate twice as far to achieve the desired flow change.

Example 2: To determine the gate setting for a desired change in water level for the current flow, for the aforementioned initial conditions, we can assume that the downstream water level does not change if the discharge does not change. The change in gate position required to raise the upstream water level by 0.2 m is found from Eq. (5-13*b*):

$$0.2\text{ m} = \frac{(1.8-1.0)}{0.5}\Delta w$$

which has for its solution $\Delta w = 0.125$ m.

5.4. CANAL POOL HYDRAULICS

Each pool is unique in its response to changes in flows through structures. Physical parameters that influence this behavior are longitudinal bed slope, cross section size and shape, length, and bed material (surface roughness). The water depth and flow rate also have their influence, as well as the types and numbers of intermediate structures, such as culverts, bridge piers, and so on.

5.4.1. Flow Conditions

Usually, at all locations along a canal pool, the water depth and velocities are different. When steady-state situations are considered in which a constant discharge runs through a pool, two flow conditions can occur:

1. *Uniform flow.* Depth is constant, so the water surface is parallel to bed slope. Note that each flow rate has a corresponding unique depth, called *normal depth.*
2. *Flow under backwater.* The downstream depth is higher than the normal depth and a more or less horizontal water surface intersects with the normal depth somewhere upstream in the pool. The higher the flow, the smaller the length of the horizontal surface. Some canals are designed such that at design capacity the water surface is at normal depth throughout the pool, while at lower flow the downstream end of the canal is under backwater. When the horizontal water surface is higher than the normal depth above the bottom level at the upstream side of the pool, the pool is completely under backwater. This is common for canals that are located along contours or for canals when the land surface slope is very flat.

Most irrigation canals are at least partially under backwater conditions for most flow conditions, where the water level at the downstream end

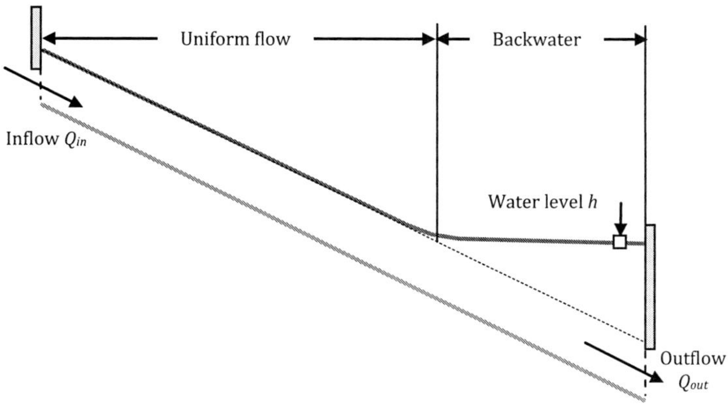

Fig. 5-1. Canal pool under backwater with uniform flow as dashed line

of the canal pool is meant to be kept at a target level, which is above normal depth (Fig. 5-1). This target level is proper for delivery of water to the offtake structures, located just upstream of the control structure. The models discussed as follows describe how the water level at the downstream end of a canal pool responds to a change in flow rate from either the upstream or downstream end. The convention used here is that upstream inflow into the canal pool and downstream outflow out from the canal pool are both positive.

Normal flow depth is often determined by use of Manning's equation (Eq. [5-15]):

$$Q = C_u \frac{AR^{2/3} S_f^{1/2}}{n} \tag{5-15}$$

where A is the cross section area; R is the hydraulic radius (area A/wetted perimeter of flow); S_f is the friction slope; n is the Manning's roughness coefficient; and C_u is the units' coefficient ($1\,\text{m}^{1/3}\,\text{s}^{-1}$ in SI units, and $1.486\,\text{ft}^{1/3}\,\text{s}^{-1}$ in English units).

Because A and R are both functions of the water depth, an iterative solution is required to determine depth from discharge. If S_f is defined as the bed slope, then the solution of Eq. (5-15) is the normal depth. If the pool is under backwater, S_f is the water surface slope. At steady flow (constant discharge), the water depths in the downstream part of the pool are determined by the settings of downstream hydraulic structures. See, for example, Eqs. (5-6), (5-8), and (5-9). Water depths at points upstream can be determined from Eq. (5-15). Calculation methods are provided in standard hydraulics textbooks (e.g., the direct step method), or calculations can be performed with steady-flow simulation models.

5.4.2. Changes in Canal Conveyance Capacity

As mentioned previously, changes in canal roughness (i.e., resistance to flow), sediment deposition, and aquatic vegetation can cause significant changes in canal water depths and volumes. These changes cause changes to the canal pool hydraulic properties. Sediment deposition is favored by low flow velocity. Canal automation can result in decreased velocity and then accelerated processes of sedimentation. For canals subject to high sediment inflow, this issue should be considered with appropriate design (using settling basins or sediment excluders) and operation.

Aquatic vegetation can consist of weeds, algae, and bryophytes. This vegetation can induce large variations of Manning's roughness coefficients and reduce drastically the canal conveyance capacity. Sedimentation can favor the development of aquatic weeds, whereas bryophytes and algae do not need a movable substratum to settle as they do not have roots.

5.4.3. Pool Volumes

Canal pools contain water while conveying it to the point of interest. When canal flow is at uniform depth, the amount of volume in the pool is dependent on the flow rate. Higher flow rates result in larger volumes. This water can be considered as being in temporary storage. It is added to the pool when the flow rate is increased, and it is removed from the pool when the flow rate is decreased. Thus, the water is delayed in arriving (and stopping) at the downstream destination. Experienced operators have observed this phenomenon, and they adjust their actions accordingly.

When the canal pool is heavily under backwater (i.e., depth well above normal depth), the volume of water within the section under backwater does not change significantly when the flow rate changes. Here, the canal pool volume primarily changes when the water level within this section changes. If the canal pool has a section under normal depth and a section under backwater, there is a transition in between where the water depth is above normal depth, but there is still a volume change (Fig. 5-2). For steep canal slopes this transition occurs very rapidly. For shallow slopes the transition can be very long and can even occur over the entire length of the pool.

Some canal operators or operational procedures explicitly compute canal pool volume changes as part of normal operations. This is discussed in Chapter 6. Here we provide procedures used to determine canal volume. The most accurate method for determining canal pool volumes is to compute the steady-flow water surface profile or backwater curve.

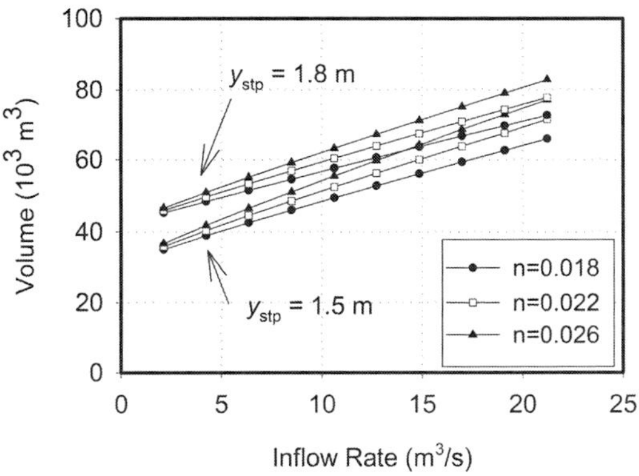

Fig. 5-2. Variation in canal pool volume as a function of discharge, Manning's n, *and downstream water level set point,* y_{stp}
Source: *Figure 4 in Bautista and Clemmens (2005); reproduced with permission*

This assumes that the canal cross section is known at key points along the canal pool and that the hydraulics of intermediate canal structures also are known. Pool volume depends on the flow rate, the downstream water level, and the flow resistance. As discussed, these volumes can be determined from steady-state backwater curves, and the results stored in tables as nomographs (e.g., from Canal de Provence, Rogier et al. 1987). Bautista and Clemmens (2005) determined pool volumes for a series of flow rates, downstream water levels, and Manning's n values and fit the data to the relationship in Eq. (5-16):

$$V = aQ^b + c \tag{5-16}$$

where the parameters a, b, and c varied with downstream water level and flow resistance, as shown in Fig. 5-2.

5.4.4. Steady and Unsteady Flow

If water flows and demands are constant, there is no need for canal automation. For large canals, operators try to minimize large flow changes and may allow small daily changes. In small canals, flow can change from zero to full capacity and back to zero in a matter of hours. The requirement for flexibility causes flow changes to be made farther and farther up the system. If flow changes are seldom and gradual, canal flows can be assumed to be under a succession of steady states. However, once significant fluctuations occur, steady-state assumptions no longer are sufficient to describe canal dynamics, and unsteady flow must be considered. This is particularly true for automated canals, where the automatic controls respond to water level changes in real time. An understanding of unsteady-state conditions can be useful for manual control, because they describe the changes in flows and water levels when going from one steady flow condition to another. These are real conditions experienced by operators, and they influence the performance of manual operations.

In the next sections, two transient effects are examined:

- Travel time of waves, or how changes in canal inflow move through a canal; this is important for routing changes in flow through the canal system so that water is available at a particular location when required.
- Response of water levels (and volumes) to upstream flow rate changes; this is important for automatic controllers that change gate settings based on observed water levels.

Travel Time of Waves There are two wave travel times of interest. The dynamic wave celerity, c_d, is defined as Eq. (5-17):

$$c_d = \sqrt{gD} \tag{5-17}$$

where D is the hydraulic depth (cross-sectional area divided by width of the water surface or top width). This is the speed of ripples in a pond. In the downstream direction, the velocity of this wave in the canal is $v + c_d$, where v is the initial average velocity of flow in the canal. This velocity can be used to predict the time when the flow first starts to increase at some location downstream. However, it is not useful by itself for predicting the time to make a flow change downstream, because a volume release takes more time to propagate than the surface ripples. When the average flow velocity is smaller than c_d, waves also move upstream. In this case, the flow is called "subcritical," which is by far the most frequent case encountered in irrigation canals.

When a flow change is made in an infinitely long canal with uniform flow (that is at normal depth), the wave propagates at a constant speed that can be predicted from continuity. The resulting equation is the kinematic wave celerity, c_{kw}, in Eq. (5-18):

$$c_{kw} = \frac{1}{B}\frac{dQ}{dy} \tag{5-18}$$

where Q is the discharge; y is the water depth; and B is the top width.

This relationship, also known as the Kleitz-Seddon formula, can give explicitly the wave celerity as a function of the canal characteristics. The kinematic wave is sometimes referred to as a "long" wave. Considering

Manning's equation and a rectangular channel of width B and normal depth y_n, one obtains Eq. (5-19):

$$c_{kw} = v_0\left(\frac{5}{3} - \frac{4}{3}\frac{y_n}{B+2y_n}\right) \tag{5-19}$$

where $v_0 = Q/(B\ y_n)$ is the mean velocity in uniform flow.

Equation (5-19) only applies to the portion of the canal not affected by the backwater curve (see Box 2).

As most canal pools are under backwater, Eq. (5-19) does not give an accurate estimation of long wave celerity, so it is not appropriate to estimate flow propagation for scheduling. The long wave celerity is rather estimated with Eq. (5-20*a*):

$$c_{lw} = \frac{L}{\Delta T} \tag{5-20a}$$

in which

$$\Delta T = \frac{dV}{dQ} \tag{5-20b}$$

is the rate of pool volume variation with respect to the flow variation. The variable L is the pool length.

The time ΔT is often used for scheduling when flow propagation must be taken into account.

Also, flow propagation results in wave attenuation, which means that rapid flow variations are attenuated during propagation. Examples of wave attenuation are provided in Box 1.

Simple Approaches to Calculate Flow Routing The water levels and flows at all locations along a canal for steady and unsteady flow are best described by the Saint Venant equations. Various modeling packages that use these equations are available for simulation of all sorts of conditions. For controller design, the Saint Venant equations are sometimes considered to be too complex. Instead, simplified models have been developed that capture the relevant dynamics. These models are described by a limited number of pool properties, which are used in the next chapter for controller design.

Some simple models can be implemented easily in a spreadsheet to calculate the response of a canal to a flow release. Among them, first-order models calculate the downstream response Q_{out} or h_d (subscript d stands for "downstream") as a function of upstream flow Q_{in}, taken a certain time earlier, due to delay, and the first-order derivative of downstream water depth or flow:

Box 1. Flow propagation process in an open channel

We consider a trapezoidal prismatic infinite channel. From an initial steady state obtained for $1\,m^3/s$, we release an extra discharge ($+0.1\,m^3/s$) over 2h. The discharge starts to increase about 10 min later 2 km downstream, but it takes about 1 h, 25 min to reach 95% of the released discharge due to attenuation. The attenuation is also responsible for the decrease of the peak discharge further in the canal (Fig. 5-3).

The dynamic wave speed is calculated for the initial steady state ($v_0 + \sqrt{gA/B} \approx 3.1\,m/s$, where B is the top width and A the cross-sectional area), whereas the mean flow velocity is $v_0 = 0.66\,m/s$, and the kinematic wave celerity is $c_{kw} = 0.89\,m/s$. The corresponding times of arrival 2 km downstream are 10 min for the short waves and 37 min for the diffusive wave.

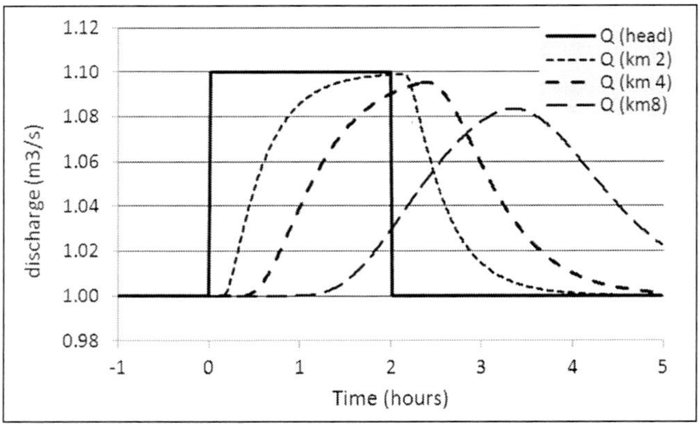

Fig. 5-3. Step release in a trapezoidal prismatic infinite channel. Bank slope = 1, bed slope = 0.5 m/km, bed width = 1 m, Manning coefficient = 0.02. The initial state corresponds to a uniform flow with 1 m^3/s

- *Integrator delay (ID) model.* The simplest description of a pool is the ID model (Schuurmans et al. 1995) that is characterized by the properties of storage area or surface area A_s (m^2) of the backwater part, and delay time τ (s). The ID model, Eq. (5-21), is written as

$$\Delta Q_{\text{out}}(t) = \Delta Q_{\text{in}}(t-\tau) - A_s \frac{dh_d(t)}{dt} \tag{5-21}$$

where h_d is the downstream water depth.

In a discrete form, taking a time step t_c, one can compute the evolution of the downstream water depth with Eq. (5-22):

$$h_d(t+t_c) = h_d(t) + \frac{Q_{\text{in}}(t-\tau)\times t_c}{A_s} - \frac{Q_{\text{out}}(t)\times t_c}{A_s} \tag{5-22}$$

In the backwater part, the water level at the downstream end of the pool, h_d (m), behaves like a storage reservoir that is influenced directly by the downstream outflow Q_{out} (m^3/s) and in a delayed manner by the upstream inflow Q_{in} (m^3/s). The values of both parameters A_s and τ change with flow rate. Lower flow rates may result in larger storage areas and usually in lower delay times. This latter relation seems contradictory as lower flow rates cause lower velocities. The reason why this does not result in higher delay times is that the major cause of delay—the length of the uniform flow part of the canal pool—decreases.

- *Integrator delay zero (IDZ) model.* The ID model describes the behavior of canal pools very accurately for nonflat pools with gradual flow changes. A sudden step in structure flow can result in an immediate step in water level that cannot be captured by the ID model. To add this behavior, a so-called zero can be added to the ID model resulting in Eq. (5-23), an IDZ model (Litrico and Fromion 2004).

$$h_d(t+t_c) = h_d(t) + \frac{Q_{\text{in}}(t-\tau)\times t_c}{A_s} - \frac{Q_{\text{out}}(t)\times t_c}{A_s} + c_{z1}\Delta Q_{\text{in}}(t-\tau) - c_{z2}\Delta Q_{\text{out}}(t) \tag{5-23}$$

where c_{z1} and c_{z2} are positive parameters that represent the immediate change in water level from a change in the delayed upstream flow and a change in the downstream flow, respectively. As for Eqs. (5-14*a*) and (5-14*b*), one can show that these parameters are of the order of $1/(B\ (c_d + v))$ and $1/(B\ (c_d - v))$.

The instantaneous effect of flow change clearly is visible when an outlet is opened, causing an immediate drop of the water level. This drop depends on the way water level is controlled.

Influence of Structures on Canal Water Level Response The characteristics of downstream structures influence the travel time of waves and how they progress through a canal. For a fixed downstream structure, the water level change resulting from a change in flow is described by Eqs. (5-6) and (5-8), for a weir and orifice, respectively. The type and characteristics of the structure then affect the water level response at the canal

Box 2. Effect of control structure on flow propagation

The same canal as described in Box 1 is considered here but with various boundary conditions at km 4. A long-crested weir or a gate controls a water level equal to $1.5y_n$, causing a 1.5-km long backwater curve in the pool. The weir is 10 m long, and the gate is 1 m wide with a fixed opening of 0.36 m. We observe a quicker response with the weir than for the uniform flow and a much slower response with the gate. At the peak discharge with the gate, the water level increases by 10 cm, and then the extra inflow is partly used to fill the pool. This does not happen with a controlled level, and then discharge is transferred downstream faster (Fig. 5-4).

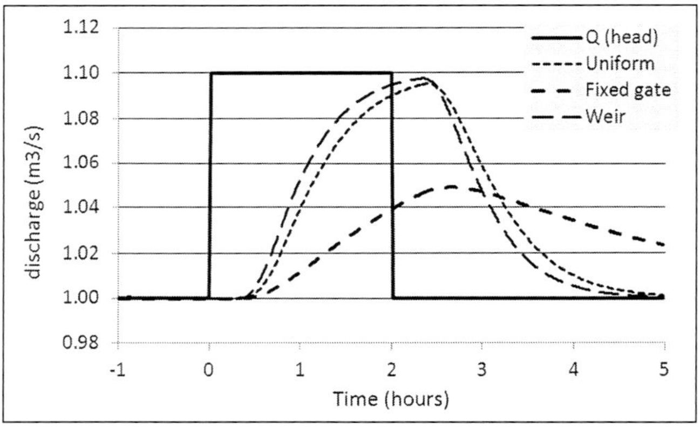

Fig. 5-4. Step response in a trapezoidal prismatic channel ended by a fixed gate, a weir, or in uniform flow

tail end, and then the storage dynamics in the pool due to backwater. An example of the relative response of different structures is shown in Box 2. Further details can be found in Strelkoff et al. (1998).

From a control perspective, it is often useful to determine the linear response of the downstream depth-discharge relationship. For a given initial depth and discharge, we can describe the relative structure response by Eq. (5-24):

$$\Delta y(t) = \Delta Q_d(t) / k_d \tag{5-24}$$

where k_d is the derivative of the gate response, as described in Eqs. (5-7) and (5-12). Under this assumption and using the IDZ model, one shows that the level response to a step withdrawal, $\Delta Q_{out,w}$, after the initial drop of $c_{z2}\Delta Q_{out,w}/(1 + k_d c_{z2})$, is Eq. (5-25):

$$\Delta y(t) = -\frac{\Delta Q_{out,w}}{k_d}\left(1 - \frac{e^{-t/K'}}{1 + k_d\, c_{z2}}\right) \tag{5-25}$$

with $K' = A_s/k_d$ time constant expressing the storage dynamics of the pool.

This will be used in Chapter 6 when we try to determine the appropriate time to change a gate setting (i.e., feedforward control).

Considering that the tail-end structure of a canal pool imposes a relationship between level variation and flow variation, h_d can be removed from Eq. (5-21). This leads to the simple first-order with delay model of Munier et al. (2010), expressed in Eq. (5-26):

$$\Delta Q_{out}(t) + K\frac{dQ_{out}(t)}{dt} = \Delta Q_{in}(t - \tau) \tag{5-26}$$

Like ID and IDZ models, this model has a delay (denoted τ), which corresponds to the delay of dynamic waves and the attenuation constant (denoted K), which takes account of storage in the pool and feedback from the downstream structure. The equivalence between Eqs. (5-26) and (5-20) can be obtained by taking $K = A_s/k_d$, but Eq. (5-26) is more generic as attenuation also occurs in the uniform part. Whereas parameters τ and K can be calculated from canal and downstream structure characteristics, they also may be obtained from field observations by making a change in flow upstream. For a step inflow of ΔQ_{in} at $t = 0$, the downstream change in discharge reduces to Eq. (5-27) (Munier et al. 2010):

$$\Delta Q_{out}(t) = \Delta Q_{in}\left(1 - e^{-(t-\tau)/K}\right) \tag{5-27}$$

For a canal at normal depth, one can also show that $\tau + K$ is close to the propagation time of the kinematic wave and Δt as defined by Eq. (5-20). An application of the model described by Eqs. (5-26) and (5-27) to real data is given in Fig. 5-5. The step change at the inlet (U/S) is $\Delta Q_{in} \approx$ 100 L/s, and the value of K is found by fitting the downstream response (D/S). In this example, the value of $K \approx$ 1 h, 40 min, computed from canal characteristics (Munier et al. 2010), is very close to the fitted value.

5.5. RESONANCE WAVES

5.5.1. Period of Resonance Waves

Canal pools that are flat, short, or deep are likely to be sensitive to resonance waves (Schuurmans 1997). These are waves that reflect against the

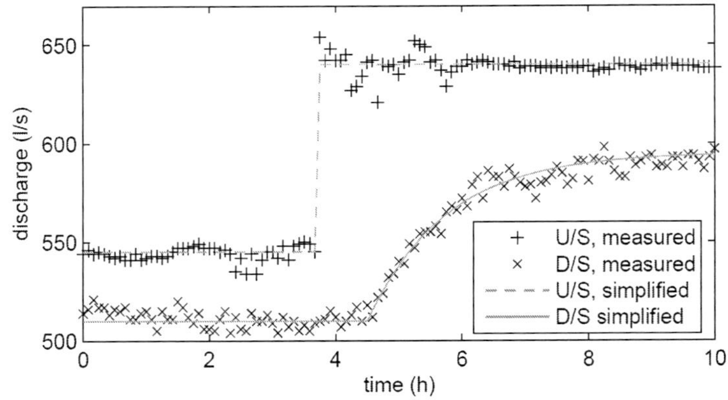

Fig. 5-5. Step response in Gignac Irrigation Canal and application of simplified transfer function Eq. (5-26) (first-order and delay), parameterized from canal characteristics
Source: *Fig. 9 in Munier et al. (2010); reproduced with permission*

upstream and downstream structures a number of times before they settle. In measurements, these waves show up after sudden changes in structure flow and are characterized by a persistent wave period. Mainly this period is dominated by the length of the pool. A rough estimation of the wave period T (s) is Eq. (5-28):

$$T = 2\frac{L}{c_d} \tag{5-28}$$

The corresponding frequency of the resonance wave is $\omega = 2\pi/T$ (rad/s).

5.5.2. Sensitivity to Resonance Waves

The resonance wave can be characterized by its sensitivity to changes in flow through the canal pool property M. This canal pool property is the magnitude of the amplitude of the sine wave in the water level divided by the amplitude of a sine wave in the flow at exactly the resonance frequency of the canal pool.

The value of M becomes higher for situations where a wave can travel more easily up and down a pool. This is the case for any situation in which the total friction that the wave experiences during its travel up and down the pool decreases; that is, with condition changes that lead to larger depths such as lower velocities and flows, smoother bed material, large backwater length, and so on. Like the other properties A_s and τ, M also changes in time with changing flows, whereas the property ω exhibits a more constant value (van Overloop 2006).

5.6. IDENTIFICATION

The hydraulic canal pool properties, A_s, τ, M, and ω, that are important for controller design (using the ID model), can be identified in field experiments or on an accurate and calibrated unsteady numerical model of the pool.

5.6.1. Identification of Integrator Delay Model Properties

The first two properties, A_s and τ, can be found with a simple step test. Whereas the water level at the downstream side is at or close to target level and the flow through the canal pool is more or less constant, a step in flow ΔQ is made at the controlling structure. The other structure's flow is kept constant. Chapter 6 describes how a constant structure flow can be imposed. In Fig. 5-6 a step test on a resonance sensitive pool is shown. Fig. 5-7 presents a step test applying an upstream flow change on a steep and long pool. For the latter situation, the auto tune variation (ATV) method can be applied to identify the storage area and delay time (see Litrico et al. 2007 for discussion).

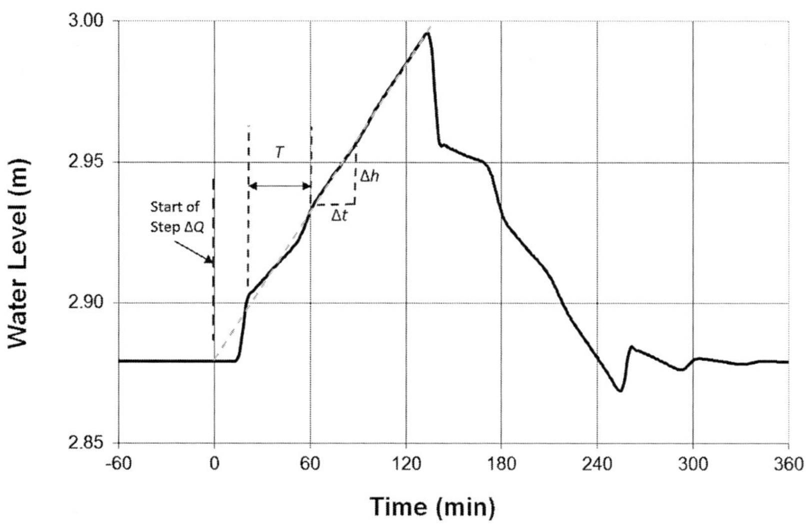

Fig. 5-6. Step test with step upward and step downward on resonance sensitive pool

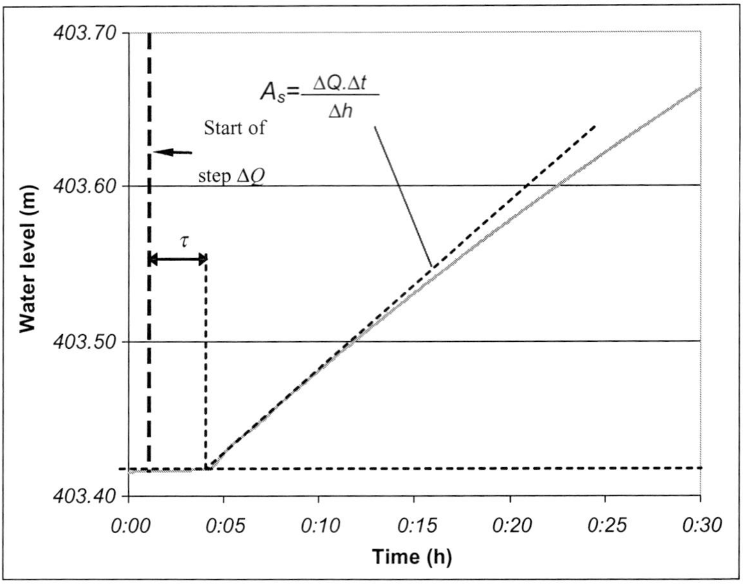

Fig. 5-7. Step test with step upward on long and steep pool

The delay time, τ, can be estimated from the time between structure flow change and start in rise of the water level. The storage area, A_s, can be estimated from the tangent of the rise line:

$$A_s = \frac{\Delta Q \Delta t}{\Delta h} \tag{5-29}$$

The start of the rise in Fig. 5-6 exhibits a resonance wave with frequency $\omega = 2\pi/T$ that dampens after some time.

5.6.2. Identification of Resonance Properties

The resonance property, M, is much harder to identify. One method is to bring the canal pool into oscillation at semi-unstable conditions using a proportional feedback controller (a PI feedback controller with integral gain set to zero; see Chapter 6 for details). By increasing the proportional gain K_p in time, the pool can be brought into a persistent oscillation, in which the amplitude of the water level is not increasing and not decreasing (Schuurmans 1997). M is then the reciprocal of this $K_{p,\text{oscillation}}$ value in Eq. (5-30):

$$M = \frac{1}{K_{p,\text{oscillation}}} \tag{5-30}$$

Obviously, it is very hard or even unacceptable to do this on an actual canal pool. Instead, an accurate validated numerical model can be used. A safer procedure is to apply system identification techniques, which involves a number of periodic small stepwise changes in the control structure flow around the estimated resonance frequency. M can be estimated in Eq. (5-31) from the amplitude of the flow signal and the resulting water level oscillation:

$$M = \frac{4}{\pi} \times \frac{\Delta h_{max}}{\Delta Q} \tag{5-31}$$

where Δh_{max} is the maximum difference between alternating tops and troughs in the oscillation of the water level around target level; and ΔQ is the amplitude of the switching flow changes.

The term $4/\pi$ is added to take the difference between a block-shaped flow signal and a sinusoidal-shaped water level signal into account. The switching flow signal can be either a so-called chirp-signal of which the frequency gradually increases from just below the estimated resonance frequency to just above this frequency or a so-called (pseudo) random binary signal with most of the frequency content closely around the estimated resonance frequency (van Overloop et al. 2010). More advanced identification techniques can be found in the extensive literature on system identification. A practical way of generating oscillations at exactly the resonance frequency is by using the auto tune variation procedure described by Clemmens et al. (2012).

5.6.3. Identification with Pool Obstructions

Objects that may complicate the identification are obstructions in the canal pool such as culverts. The most important rule to be able to apply the aforementioned identification methods is to apply the test signals (step or oscillations) with the structure that is intended to control the water level in that particular pool.

5.6.4. Influence of Flow Conditions on Identified Properties

A main issue is that each set of identified canal pool properties is valid for a certain flow condition. Now the most important sets that need to be available when going toward controller design are those for minimum flow (not zero) and maximum flow. An exception to this involves the resonance properties. These only need to be known at minimum flow. Often, especially for new canals, only the flow design capacity of a pool is known. In that case, the low flow condition can be taken at 10% of the design capacity. After the identification phase, we assume that we have attained the following properties: $A_{s,\mathrm{low}}$, $A_{s,\mathrm{high}}$, $\tau_{s,\mathrm{low}}$, $\tau_{s,\mathrm{high}}$, M, and ω, where the latter two need to be identified at low flow.

5.7. REFERENCES

Aisenbrey, A. J., Jr., Hayes, R. B., Warren, H. J., Winsett, D. L., and Young, R. B. (1978). *Design of small canal structures*. U.S. Department of Interior, Bureau of Reclamation, Denver.

Bautista, E., and Clemmens, A. J. (2005). "Volume compensation method for routing irrigation canal demand changes." *J. Irrig. Drain. Eng.*, 131(6), 494–503.

Clement, R. (1965). "Computation of flows in irrigation systems operating 'on demand'." *La Societe du Canal de Provence et d'Amenagement de la Region Provencale*, Le Tholonet. Trans. U.S. Bureau of Reclamation, Denver.

Clemmens, A. J. (1986). "Canal capacities for demand under surface irrigation." *J. Irrig. Drain. Eng.*, 112(4), 331–347.

Clemmens, A. J. (1987). "Arranged delivery schedules." *Planning, operation, rehabilitation and automation of irrigation water delivery systems*, D. D. Zimbelman, ed., ASCE, New York, 57–67.

Clemmens, A. J., Wahl, T. L., Bos, M. G., and Replogle, J. A. (2001). *Water measurement with flumes and weirs*, ILRI Publication 58, International Institute for Land Reclamation and Improvement, Wageningen, The Netherlands.

Clemmens, A. J., Litrico, X., van Overloop, P.-J., and Strand, R. J. (2012). "Estimating canal pool resonance with auto tune variation." *J. Irrig. Drain. Eng.*, 138(1), 9–15.

Litrico, X., and Fromion, V. (2004). "Simplified modeling of irrigation canals for controller design." *J. Irrig. Drain. Eng.*, 130(5), 373–383.

Litrico, X., Malaterre, P.-O., Baume, J.-P., Vion, P.-Y., and Ribot-Bruno, J. (2007). "Automatic tuning of PI controllers for an irrigation canal pool." *J. Irrig. Drain. Eng.*, 133(1), 27–37.

Malaterre, P.-O., and Baume, J.-P. (1997). "Optimum choice of control action variables and linked algorithms: Comparison of different alternatives." *ASCE-ICID Workshop of Modernization of Irrigation Water Delivery Systems*, Phoenix, AZ.

Munier, S., Belaud, G., and Litrico, X. (2010). "Closed-form expression of the response time of an open channel." *J. Irrig. Drain. Eng.*, 136(10), 677–684.

Rogier, D., Coeuret, C., and Brémond, J. (1987). "Dynamic regulation on the Canal de Provence." *Planning, operation, rehabilitation and automation of irrigation water delivery systems*, D. D. Zimbelman, ed., ASCE, New York, 180–200.

Schuurmans, J. (1997). "Control of water levels in open channels." Ph.D. dissertation, Delft University of Technology, Delft, The Netherlands.

Schuurmans, J., Bosgra, O. H., and Brouwer, R. (1995). "Open-channel flow model approximation for controller design." *Appl. Math. Model.*, 19(9), 525–530.

Strelkoff, T. S., Deltour, J. L., Burt, C. M., Clemmens, A. J., and Baume, J. P. (1998). "Influence of canal geometry and dynamics on controllability." *J. Irrig. Drain. Eng.*, 124(1), 16–22.

U.S. Department of Agriculture, Soil Conservation Service. (1977). *Design of Open Channels*, Technical Release No. 25, U.S. Department of Agriculture, Soil Conservation Service, Washington, DC.

van Overloop, P. J. (2006). "Model predictive control on open water systems." Ph.D. dissertation, Delft University of Technology, Delft, The Netherlands.

van Overloop, P. J., Miltenburg, I. J., Bombois, X., Clemmens, A. J., Strand, R. J., van de Giesen, N. C., et al. (2010). "Identification of resonance waves in open water channels." *Control Eng. Pract.*, 18(8), 863–872.

CHAPTER 6
CONTROL METHODS

6.1. INTRODUCTION

Various concepts for the control of irrigation canals are described in previous chapters: Chapter 2 discussed control with mechanical structures, and Chapter 4 described supply-and-demand-oriented control strategies. These methods are used routinely by canal operators to operate canals manually. The purpose of this chapter is to present methods by which these control methods can be implemented automatically via electronic devices such as PLCs, RTUs, computers, and other devices. The intent of canal automation is to improve the operation of the water distribution system, which typically means better service to farmers; canal automation is intended to improve some aspect of operations by performing controls that would be difficult to do manually.

An effective automatic control method must have well-defined inputs and outputs. Where possible, these inputs and outputs should be the same as those used by canal operators. In some cases, the operator has more information than is available through the automatic device. In such cases, operation can be a combination of manual and automatic control. In addition, most successful canal automation implementations phase in automatic controls gradually to ensure that they function properly and provide the desired outcomes.

Consider a canal with turnouts that have water measurement devices that are not interfaced with the control device. The operator adjusts the turnout gate to provide the required flow rate based on the water measurement device. The automatic control device controls the water level in the canal so that the outlet has a constant upstream head. By maintaining this water level, the automatic device essentially maintains a constant turnout flow (assuming no downstream influences). Thus, the overall control strategy is a mix of manual and automatic control features, as current canal automation allows. As technology improves over time, one might expect less manual control and more automatic control.

In determining which features to automate, one must consider how the necessary information will be provided to the automatic control device. A canal headgate can provide automatic control of inflow to a canal, but the required flow somehow must be sent to the control routine. A common procedure is for the operator to provide this information as an input. However, if demands from downstream users are entered into a water ordering program (i.e., from an arranged demand schedule), one might envision that these data could then be used to control canal inflow. Although this is feasible, there is a certain amount of danger in automating data directly from a water ordering program (e.g., feedforward only). If demand changes are not properly updated, this could cause disastrous results. In addition, flow rate measurement errors and flow adjustment errors also have to be taken into account. Such adjustments can be made either by the canal operator or through feedback control, as discussed later.

Throughout this chapter the word "controller" is used to refer to the control algorithms that calculate the control actions irrespective of the device on which these run.

6.1.1. Controlled Variables

The *controlled variables* are those variables that the automatic control system attempts to maintain at a constant value, typically called the *set point*. For irrigation canals, these are usually flow rate, canal water level, and pipeline pressure. This manual does not deal with pipeline pressure control, although many of the control concepts are similar. The control scheme may attempt to control different variables at different locations in the system. The general requirement for irrigation canal systems is to provide the correct flow rate to users. Operation of large canals often must control fluctuations in water levels to protect the canal lining. Common actions for lateral canals is to control the flow rate at the head of each canal and to control the water level at downstream check structures in an attempt to maintain constant pressure for the outlet. Therefore, careful consideration must be used when deciding which flow rates and water levels to control when designing a canal control scheme.

6.1.2. Control-Action Variables

The *control-action variables* are the variables that the control system manipulates to bring the controlled variable back to its set point.

Once the controlled variable is determined, an appropriate control-action variable must be selected. For canals, the control-action variable is almost always a gate or valve position (or, for pump inflow and outflows, on and off or variable frequency drive (VFD) frequency), even though the actual controller output might be a flow rate change. For an automatic water level controller, for example, the water level is the controlled variable and the gate position is the control-action variable. Thus, the controller moves the gate to bring the water level back to its set point. In another situation, the water level in a pool can be influenced by changing the gate position of the check gate that is either downstream or upstream of the water level measurement. Either gate can be used to change the water level in the pool, but the results of using one or the other have radically different net results on operations (see Chapter 4 for details on upstream or downstream control).

6.2. IMPLEMENTATION OPTIONS

Automatic controls can be organized in a variety of ways, just as there are a variety of options for automatic versus manual control:

- *Local Control*: Automatic control at individual structures is often called *local control* and can be classified as a *single-input, single-output* (SISO) control method. Commonly, the control instrumentation is located at the structure site, although many SCADA systems also have built-in automatic control functions, which can be used to implement local control actions. However, setting up local control in this centralized fashion generally is not recommended due to a history of communication timing issues and communication bandwidth issues.
- *Centralized Control*: Control decisions made at a central location, such as at the SCADA node, is often called *centralized control*, particularly if the control decision for one location is based on observations from multiple locations. Often all relevant observations along the canal are transferred to a central control room, where the information is processed and control actions to all control structures are sent out over appropriate communication lines. One advantage of centralized control is that the operator/controller can observe all water levels simultaneously. As such, central controllers are considered *multi-input, multi-output* (MIMO) controllers. Application of SISO controllers from a central location is not considered centralized control.
- *Hierarchical Control*: If the set point for a local controller is determined from a central controller, then the control is considered

hierarchical. This is becoming the most common form of control for canal operations. The decoupling technique described in the next section is an example of hierarchical control.

- *Distributed Coordination Control*: Local controllers communicating among themselves with neighboring local controllers are referred to as *distributed coordination control*. This might be applied to adjacent structures along a canal, where a controller at one site receives information from controllers immediately upstream and downstream. Typically this is not used on irrigation canals.

6.3. DECOUPLING POOLS AND STRUCTURES

In Chapter 5, canal pool properties were defined in terms of canal flow rate. Controller design uses these properties in linear models to design control constants. Yet, the control actions are often changes in gate position. For a given flow, the required gate position is a nonlinear function of upstream and downstream water level. Check structures often have multiple gates and weirs, with gates sometimes of different widths. Yet, regardless of the check structure configuration, the canal water level response is based on flow rate change. To simplify the design of controllers (which are usually linear) and to simplify implementation, most canal controllers determine the flow rate change for a check structure, and the gate position change required to achieve the prescribed flow rate is computed separately.

The local controller needs to determine the change in gate position, Δh_g, for a desired change in gate discharge, ΔQ. For given upstream and downstream levels, this is approximately linear. The change in gate opening then can be determined from Eq. (6-1):

$$\Delta h_g = \Delta Q \times \delta h_g / \delta Q \tag{6-1}$$

where $\delta h_g / \delta Q$ may change over the range of upstream and downstream levels. The ratio $\delta h_g / \delta Q$ can be determined from the head discharge equation. If the relationship is truly linear, then $\Delta h_g = \Delta Q \times h_g / Q$.

6.4. ROUTING DEMAND CHANGES THROUGH A CANAL

6.4.1. Approach

Routing water through a canal can be done by using the delay times from the headgate to turnouts. The simplest approach is to record the delay times suggested by the operators and to implement feedforward control based on those times. This does not consider how the delay times differ at different flow rates or with different percentage flow changes.

For some canals, this effect is minor, whereas for others it can be significant. Bautista and Clemmens (2005) developed a simple scheme for routing flow changes based on the change in canal volume resulting from a change in canal flow. The delay time ΔT is found from Eq. (6-2):

$$\Delta T = \Delta V / \Delta Q \tag{6-2}$$

where ΔV is the change in volume in going from one steady-state flow to another, and ΔQ is the change in steady-state flow (see Chapter 5, Eq. [5-20]).

The canal pool volumes are calculated using Eq. (6-3) for the different canal flow rates so that

$$\Delta V = V\,(Q_{\mathrm{in}} + \Delta Q) - V(Q_{\mathrm{in}}) \tag{6-3}$$

where Q_{in} is the initial canal flow.

This requires knowledge of the canal pool volumes versus time, which requires a detailed analysis of these canal pools. The pool volumes are determined for a series of flow rates, and the data are fitted to the following relationship for easy implementation using Eqs. (6-4) and (6-5), where a, b, and c are constant coefficients valid for an individual canal pool:

$$V = aQ_{\mathrm{in}}^{b} + c \tag{6-4}$$

and then

$$\Delta T = abQ_{\mathrm{in}}^{b-1} \tag{6-5}$$

This method has two common complications:

1. Use of this approach requires knowledge of the flow rate in all pools at all times. In practice, one can use the sum of the downstream demands for those pools or the intended flow for those pools. This is sufficiently accurate in most cases, but it does require that the water orders are part of the automation scheme. In many cases, these records are entered in water ordering and billing. The drawback is that this information often is not available for automatic control. It is provided to operators in one form or another but often not used as an input to the SCADA system. As a result, when changes are made by operators in the field, these records are not updated automatically. Even if the records were provided to the SCADA or automation software, effort would have to be made to keep these up to date. Implementing delivery routing without accurate information on current orders could result in poor performance.
2. A considerable amount of effort is required to develop the volume relationships described by Eq. (6-4). If the water in the pool is at

normal depth, calculation of volume is relatively simple, although it is a function of channel roughness (e.g., Manning's n). In addition, the volume depends on the amount of backwater, which is influenced by the downstream depth. Thus, the coefficients a, b, and c in Eq. (6-5) are all functions of Manning's n and depth. Calculations can assume that the water level is at the downstream set point, and the Manning's n value can be specified to account for seasonal changes in resistance.

The volume compensation method for routing flow changes assumes a change from one steady state to another and that the water level at the downstream end of the pool is at the set point for both of these steady states. Given the effort required to come up with the volume-compensation coefficients, calculation of these delay times may be considered. The concept here is to determine the time to change the outlet so that the downstream supply provides the correct volume. Because of wave dispersion, it is not possible to maintain the correct flow rates through the transition, but one can provide the correct volume. Chapter 5 provides equations that describe how flow rate changes propagate through a canal pool. If the flow change at the downstream end of the pool is integrated over time, one can determine how the water level and downstream flow change over time.

For an outlet with a fixed discharge, such as a pump, the response of the downstream water level based on a fixed withdrawal can be found from the IDZ model as in Eq. (6-6):

$$q_{\text{out},d}(t) = -q_{\text{out},w}(0)\left(1 - \frac{e^{-(t-T_w)/K'}}{1+k_d c_{z2}}\right) \tag{6-6}$$

where $q_{\text{out},d}(t)$is the downstream flow change at time t; $q_{\text{out},w}(0)$ is flow change when the outlet is first opened; T_w is the ideal time to open the outlet; K' is the attenuation resulting from a downstream discharge change (see Chapter 5, Eq. [5-24]); c_{z2} is the parameter that takes account of instantaneous water level drop caused by opening the outlet; and k_d is the rate of change of check gate discharge with water level, namely $k_d = dQ_{\text{out},d}/dy$ for the check structure (subscript d) (see Chapter 5, Eq. [5-23]).

Integrating the resulting volumes over time from 0 to T_w (before the outlet is opened) and from T_w to infinity (after the outlet is opened) gives the following solutions for T_w in Eq. (6-7):

$$T_w = \tau + t_w = \tau + K - \frac{K'}{1+k_d c_{z2}} \tag{6-7}$$

where t_w is the time to open the outlet after the short wave has arrived (described by τ).

This means that the time to change the offtake to provide volume compensation depends on the hydraulics of the downstream structure (check structure). If the downstream level is held constant, $k_d \rightarrow \infty$, and we get Eq. (6-8):

$$T_w = \tau + K \tag{6-8}$$

In this case, the volume compensation time is the rise time of the system. The rise time of the system occurs when the term in the parentheses is $(1 - 1/e) \approx 63\%$ of the total flow response time. Eq. (6-8) is the time of travel of long waves (see Chapter 5). Thus, Eq. (6-7) shows that the time to open the outlet falls between the travel time of the dynamic wave (short) and the kinematic wave (long).

For conditions where the outlet is operated by gravity and the check gate is a constant flow device, the result is the same. When the outlet and the check gate both are operated by gravity, the volume compensation time is shorter than the time given by Eq. (6-8). This time difference is small when the check gate is much larger than the turnout gate. When they are of equal size, the volume compensation time is the minimum. For outlets of equal size, the value of t_w is changed by less than 10% from the value computed previously. An approximate solution is found in Belaud et al. (2013).

6.4.2. Routing Schedule Example

This example does not deal with the calculation of delay times but instead deals with the application of these delay times under operating conditions. When developing a schedule (Box 1), operators often group adds (extra water flow required) and cuts (reductions of water flow) within a pool to reduce the number of changes at each control structure. For larger canals, these adds and cuts may be located somewhat distant from each other, but if they are within the same pool, the pool volume can act as a buffer, and the resulting set point changes stay within acceptable tolerance. Grouping also may occur across time, allowing for a single change to be made for delivery changes that are to occur at different times. Rules for grouping cuts and adds are derived from experience and may be ignored altogether if they are very small in comparison to the capacity of the canal pool.

The schedule is developed by summing the changes in flow from downstream to upstream using delay times for each pool from the point of delivery to the head. Each change can be routed separately. Changes are additive: No adjustments are made based on interaction between flow changes.

Box 1. Manual schedule for a downstream set point control canal

A canal consisting of six pools having between one and three turnouts per pool has the schedule shown in Table 6-1.

The first step is to look at the schedule and see which changes can be combined or ignored.

1. Entries 1 and 2 occur at the same turnout 2.5 h apart and exactly offset one another. Experience shows that they can be ignored. This is justified by the fact that the corresponding volume is much smaller than the pool volume.
2. Entries 5 and 6 exactly offset one another but occur more than 7 h apart. These will be kept separate.
3. Entries 9, 14, and 15 all occur within the first hours of the day and are in the same pool. They can be combined into one add of 0.015 at 0:30.

Table 6-1. Example Schedule

Entry	Pool	Order point	Existing delivery flow rate (m^3/s)	Time	Add/cut (m^3/s)
1	1	1 C SRPMIC Large Gate	1.600	6:00	0.200
2				8:30	–0.200
3	1	1 B SRPMIC Small Gate			
4	1	1 A SRPMIC Pumping Station	1.280	6:00	0.640
5	2	1 Laterals 010–019	0.055	6:30	0.060
6				13:45	–0.060
7	3	1 Laterals 020–030	0.085	7:45	–0.050
8	5	1 Laterals 035–050	0.130	0:01	0.010
9	6	1 Laterals 060–064	0.300	1:15	–0.060
10				2:00	0.025
11				5:15	0.085
12				6:55	–0.085
13				20:15	–0.085
14	6	1 Laterals 061–080	0.131	0:00	0.025
15				0:00	0.050
16				2:30	0.065
17				5:15	0.025
18				7:00	–0.110
19				9:45	–0.025
20				14:30	–0.065

4. Entries 10 and 16 occur within 0.5h of each other and can be combined.
5. Entries 11 and 12 exactly offset each other and are less than 2h apart. They can be ignored.
6. Entries 17 and 19 exactly offset one another but are 4.5h apart. The experience of the water master will determine whether to keep these in the schedule. They will stay for the purpose of this example.
7. Entries 3, 4, 7, 8, 13, 18, 19, and 20 remain unchanged.

Each change in delivery is routed upstream through the pools using a constant average delay time for each pool.

- Use delay times for each pool determined by experience. Here we assume 4 hours in pool 1, 3h in pool 2, 1 hour in pool 3, 0.5h in pool 4, and 3h in pool 5.
- Add delay times together to determine when to change control structure setting. For example, entry 13 (cut of 0.085) would be scheduled:
  - Pool 1 = 20:15 – (4+3+1+0.5+3) = 8:45
  - Pool 2 = 20:15 – (3+1+0.5+3) = 12:45
  - Pool 3 = 20:15 – (1+0.5+3) = 15:45
  - Pool 4 = 20:15 – (0.5+3) = 16:45
  - Pool 5 = 20:15 – 3 = 17:15
  - Pool 6 = 20:15

Often a change scheduled for early in the day at the downstream end of the canal requires a change before the beginning of the day to meet the requirements of the delay times. In such cases, we set those changes to occur at the beginning of the day. The scheduled changes for the control structure at the head of each pool (m^3/s) are shown in Table 6-2.

This schedule can be further refined by combining changes as in the first step. Sometimes schedulers add in a little extra when adding to a delivery or take away a little more when cutting a delivery to make up for pool volume change. These usually are based on rules developed from experience. Changes to make up for pool volume change may be ignored altogether and made up later in the day when water levels drift away from the set point.

Table 6-2. Schedule of Check Gate Flow Changes

Pool	1	Pool	2	Pool	3	Pool	4	Pool	5	Pool	6
00:00	0.005	00:00	0.030	00:00	0.115	00:00	0.115	00:00	0.115	0:30	0.015
0:45	–0.050	02:15	–0.025	00:45	0.025	01:45	0.025	02:15	0.025	2:00	0.090
02:30	0.060	04:45	–0.050	02:30	–0.110	03:30	–0.110	04:00	–0.110	5:15	0.025
03:00	–0.065	06:30	0.060	05:15	–0.025	06:15	–0.025	06:45	–0.025	7:00	–0.110
06:00	0.640	07:00	–0.065	07:45	–0.050	11:00	–0.065	11:30	–0.065	9:45	–0.025
8:45	–0.085	12:45	–0.085	10:00	–0.065	16:45	–0.085	17:15	–0.085	14:30	–0.065
9:45	–0.060	13:45	–0.060	15:45	–0.085					20:15	–0.085

Notes: Changes are at the upstream end of each pool. Changes 10 and 16 are scheduled together.

Routing Unscheduled Deliveries If a flow change has not been scheduled ahead of time, the procedures used here can provide the time that the outlet is opened, given the initial time water is released.

During the start of the irrigation season, some canals (and rivers where water is released from a dam) have unpredictable response times until they have been wetted for the first time during the season. Here the water cannot be delivered with a schedule, and the water is released to the outlet when the flow change arrives. If the response cannot be predicted, then the process should not be automated.

6.4.3. Accounting for Routing (Feedforward Control) Errors

In most irrigation canals, flow rate measurements (and controls) are not sufficiently accurate to ensure that the rate of flow entering the canal is equal to the rate of flow required downstream. There are always mismatches. If canal operators do not make adjustments to create this balance, some users either receive not enough or too much water or the canal will spill. If the initial flow rates are reasonably accurate, only a few adjustments may be needed. The most common approaches to correcting this kind of problem are as follows:

- Improve the flow measurement devices used.
- Hire labor to provide downstream feedback control manually.
- Use automatic feedback control.
- Overdeliver and collect the spilled water downstream.

Whether feedback control is done manually or automatically, routing (sometimes called feedforward control) and feedback control generally are done simultaneously. In both cases, control is incremental, that is, provides a change from the current condition. Use of incremental control allows feedforward and feedback signals to be superimposed.

6.5. FEEDBACK CONTROL

This section describes the mathematical procedures needed to tune controllers for the control of water levels, flow rates, or volumes within a canal. These methods use measurements of water level or flow rate to determine what control actions can bring the water level, flow rate, or canal pool volume to a set point value (see Chapter 4 for a discussion of basic control concepts). Canal pool volume generally is based on measurements of water level; thus, volume control can be thought of as a special version of water level control and not treated separately.

Tuning feedback control is more straightforward when the canal pool hydraulics are decoupled from the structure hydraulics as described previously. Water level response generally is based on flow changes that occur in the pool (see Chapter 5). Therefore, to control a water level, the automatic control must change the flow rate into or out of the pool. A separate calculation in a local controller/algorithm is made to determine the change in gate or valve position to implement this control.

6.5.1. Filtering

Filtering of measurement signals is advisable to remove noise and other high-frequency content that is not relevant for control.

First-Order, Low-Pass Filter A simple filter algorithm that can be used is a *first-order, low-pass filter,* in this case applied to a water level measurement with Eq. (6-9):

$$h_f(t) = c_f h_f(t - t_c) + (1 - c_f)h(t) \tag{6-9}$$

where $h_f(t)$ (m) is the new filtered water level; $h_f(t - t_c)$ (m) is the filtered water level at the previous control time step; and c_f is the filter constant, a value between 0 and 1.

To avoid so-called aliasing, where nonexisting periodical oscillations appear in the signal due to a too low sampling frequency, this filter is, in fact, always needed with a value of 0.667 or higher.

Resonance Wave Filtering Resonance waves can be present in canal pools, especially in flat and short pools where waves can travel forth and back easily through the canal, reflecting on the ends of the pool. It is not possible to eliminate these waves actively using control, so instead they should be ignored using appropriate filtering. A value for c_f that dampens the resonance wave in the signal considerably and sufficiently can be calculated using Eq. (6-10):

$$c_f = e^{\left(\frac{-t_c}{\sqrt{\frac{A_s M}{\omega}}}\right)} \tag{6-10}$$

where A_s is the average storage area of the backwater part; M is the resonance magnitude; and ω is the resonance frequency (see Chapter 5 for an explanation of these variables and how values are estimated).

6.5.2. Control Direction

Feedback control can be applied with two different control directions:

- *Downstream* control is where the water level at the downstream end of the pool is controlled by the structure at the upstream end of the pool.
- *Upstream* control is where the water level at the downstream end of the pool is controlled by the structure at the downstream end of the pool.

The control direction is very important and dictates the overall control strategy. When focusing on the pool, the definition of the flow variables that determine the behavior become different for the two cases. With the integrator delay (ID) model, the canal response is described by Eq. (6-11) (see Eq. [5-22] in Chapter 5):

$$h_d(t+t_c) = h_d(t) + \frac{Q_{\text{in}}(t-\tau)\times t_c}{A_s} - \frac{Q_{\text{out},d}(t)\times t_c}{A_s} - \frac{Q_{\text{out},w}(t)\times t_c}{A_s} \quad (6\text{-}11)$$

where h_d is the water level (relative to a fixed reference) at the downstream end of the pool; t is time; τ is the pool response time; t_c is the control time step; Q_{in} is the inflow at the upstream end; $Q_{\text{out},d}$ is the outflow at the downstream end; $Q_{\text{out},w}$ is the flow to an outlet or branch canal.

With upstream control, the controlled flow, Q_c, is $Q_{\text{out},d}$, and the primary disturbances are Q_{in} and $Q_{\text{out},w}$. With downstream control, the controlled flow, Q_c, is Q_{in}, and the primary disturbances are $Q_{\text{out},d}$ and $Q_{\text{out},w}$. This is important to understand, because feedback control uses the water level deviation to modify the controlled flow, whereas feedforward control uses the (anticipated) disturbance flow to minimize the effect that the disturbance flow has on the water level. It should be noted that feedforward control is never perfect because of wave dispersion and other model mismatches. Thus, feedforward control, when used in combination with feedback control, causes a disturbance on the controlled flow, as well as what is normally considered the disturbance flow.

6.5.3. Proportional Integral Control Algorithms for Water Levels

This section describes the commonly used proportional integral (PI) controller, including an example of how to tune a feedback controller. The PI-controller is a PID-controller where the derivative gain is set to zero. It gives a feedback signal as soon as the water level deviates from water level reference. This target level or set point usually is constant and chosen to give enough head and backwater storage for the offtake structures that are close to the location of the water level measurement. First, the deviation or error e (m) is calculated using Eq. (6-12):

$$e(t) = h_f(t) - h_{\text{ref}} \tag{6-12}$$

where h_f (m) is the filtered measured water level; and h_{ref} (m) is the reference water level. Note that the measured water level first needs to be filtered, either for anti-aliasing or for attenuating possible resonance waves (see Chapter 5 for details).

Next, the required control action is calculated in terms of flow. It is strongly advised to use the difference version of the PI-controller, as it avoids problems with wind-up of the integral term. The difference version of the PI-controller can be seen as the derivative of the normal PI-control formulation in which the flow is a function of the error and the summation of the error. To use the difference PI-controller, the present flow $Q_c(t)$ needs to be measured or estimated, and the following PI-algorithm needs to be calculated with Eq. (6-13):

$$\Delta Q_c(t) = K_P(e(t) - e(t - t_c)) + K_I e(t - t_c) \tag{6-13}$$

where $\Delta Q_c(t)$ (m^3/s) is the required change in flow; $e(t - t_c)$ is the error at the previous control time step; K_P is the proportional gain; and K_I is the integral gain. The values for the latter two can be determined with the tuning rules that are described as follows.

The following procedure (from Schuurmans 1997) can be used to find proper values for K_P and K_I using the canal pool properties described in the previous chapter.

- Using Eqs. (6-14) and (6-15), calculate two possible K_P values—one for canal pools for which the tuning is restricted by the delay time $K_{P,\text{delay}}$ and one for pools for which the tuning is restricted by the resonance $K_{P,\text{resonance}}$:

$$K_{P,\text{delay}} = \frac{\min(A_{s,\text{low}}, A_{s,\text{high}})}{3 \times \max(\tau_{\text{low}}, \tau_{\text{high}})} \tag{6-14}$$

$$K_{P,\text{resonance}} = \frac{1}{2}\sqrt{\frac{A_s \omega}{M}} \tag{6-15}$$

  where $A_{s,\text{low}}$ and $A_{s,\text{high}}$ are the storage area of the backwater part at low and high flow conditions, respectively; τ_{low} and τ_{high} are the delay times at low and high flow conditions, respectively; A_s is the average storage area of the backwater part ($[A_{s,\text{low}} + A_{s,\text{high}}]/2$);
  M is the resonance magnitude; and ω is the resonance frequency (see Chapter 5 for an explanation of these variables and how values are estimated).

- When $K_{P,\text{delay}} \leq K_{P,\text{resonance}}$ this means the tuning is restricted by the delay time, so use Eq. (6-16):

$$K_P = \frac{\min(A_{s,\text{low}}, A_{s,\text{high}})}{3\times\max(\tau_{\text{low}}, \tau_{\text{high}})} \text{ and } K_I = \frac{\min(A_{s,\text{low}}, A_{s,\text{high}})\times t_c}{18\times\max(\tau_{\text{low}}^2, \tau_{\text{high}}^2)} \tag{6-16}$$

  where t_c is the control time step.
- When $K_{P,\text{resonance}} < K_{P,\text{delay}}$ this means the tuning is restricted by the resonances, so use Eq. (6-17):

$$c_f = e^{\left(\frac{-t_c}{\sqrt{\frac{A_s M}{\omega}}}\right)}, \; K_{P,\text{resonance}} = \frac{1}{2}\sqrt{\frac{A_s\omega}{M}} \text{ and } K_I = \frac{t_c\omega}{12M} \tag{6-17}$$

  where c_f is the filter coefficient of the first-order, low-pass filter that needs to be used to attenuate the resonance waves (see Chapter 5 for details). Note that if the delay time is small (e.g., for short pools), $K_{P,\text{resonance}}$ will be smaller than $K_{P,\text{delay}}$, and so this resonance restricted control gain needs to be used.

In these tuning rules, the sign is positive. This is valid for upstream control. For downstream control, a negative sign needs to be added to both the values of K_P and K_I. These control coefficients are valid only for control of a single gate. Methods to control multiple gates are discussed in a later section.

6.5.4. Control Time Step Selection

Regarding the selection of the control time step, t_c, the following remarks can be made. The smaller the time step, the higher the performance of the control can become. Conversely, sensors, communications, algorithms, PLCs, and others may not allow a very small time step. In general, for irrigation and small drainage canal pools, a control time step in the range of 1 to 10 min is a good choice. Published literature on control theory suggests determining the time step as 10% of the time constant of the system that is controlled (Van de Vegte 1990). This time constant can be estimated by the time it takes after a step change on the input until the output reaches 63% of its end value. Too-frequent movements can cause wear and tear on the structures, so the number of structure changes can be reduced by implementing a deadband on the output of the controller and an accumulator in the controller logic. For example, when the required change in structure setting Δh_g is lower than a threshold $\Delta h_{g,\min}$ (e.g., 1 cm), the gate is not moved. Instead, the change in structure flow is kept in the controller's memory and added to the next required structure change.

When the time step is large, it can add too much delay to the control loop that is tuned with the aforementioned tuning rules for PI-control. Check Eqs. (6-18) and (6-19):

- For the delay time restricted feedback control, if $\frac{t_c}{\max(\tau_{\text{low}}, \tau_{\text{high}})} > 0.4091$, then use:

$$K_P = 0.136 \times \frac{\min(A_{s,\text{low}}, A_{s,\text{high}})}{t_c}, \; K_I = 0.0093 \times \frac{\min(A_{s,\text{low}}, A_{s,\text{high}})}{t_c} \quad (6\text{-}18)$$

- For the resonance restricted feedback control, if $\frac{t_c}{M \times A_s} > 0.3$, then use:

$$c_f = e^{-0.3} = 0.741, \; K_P = 0.15 \times \frac{A_s}{t_c}, \; K_I = 0.0075 \times \frac{A_s}{t_c} \quad (6\text{-}19)$$

6.5.5. Disturbance Amplification

The PI-controller is an algorithm that uses a single input and generates a single output, which is referred to as a SISO control. Because canals have multiple pools in series, when gate controls are tuned individually, the operation of the entire canal will exhibit significant oscillation due to the phenomenon called *disturbance amplification*. One pool is controlled by its control structure, which generates control actions that are proper for its own pool but at the same time creates a disturbance for the next pool. So, this next pool not only has to take care of its own deviations but also the created disturbance from the adjacent pool. For the next pool downstream, it is even worse, and so on all along the entire canal. This can be avoided by using

- A MIMO controller, such as a linear quadratic regulator or H_∞, in which an optimization calculates the feedback matrix, linking all water level deviations to all structure flows of the entire canal (Malaterre 1995; Litrico and Fromion 2009);
- A centralized tuning of all separate PI–controllers taking the entire canal into account, also by means of optimization (van Overloop et al. 2005); the result of this optimization is usually a gradual reduction of the K_I values of the PI–controllers in series; and
- Decoupling of the canal pools.

When a feedback controller makes changes to the control flow to correct for a water level deviation from target level, for example, caused by a change in offtake flow, it is clearly beneficial for that pool. However, as pools are linked along a canal, the change in control flow disturbs the situation in the adjacent pool, causing the water level in that pool to start deviating from the target level. For this adjacent pool, this can be seen as a *disturbance flow* that is generally known, as it is calculated by the other pool's controller. When looking at the ID model, the least deviation from target level occurs when the known disturbance flow is communicated directly to the other side of the pool, and the control flow there is set to this value. When all pools implement this mechanism, all controllers communicate this disturbance all the way across the canal either upstream, in case of downstream control direction, or all the way downstream, in case of upstream control direction. This setup is referred to as *decoupling pools* because a water level deviation in one pool will, theoretically, not influence the deviations in other pools, even though the transported flow through these pools can change considerably. Controller performance often improves when the decoupling effect is reduced by using a factor of less than 100% on the decoupling flow sent to the next control structure.

6.5.6. Linear Quadratic Regulator Control

Linear quadratic regulator (LQR) control requires a model of the entire canal consisting of all water level deviations and control flow interactions (Malaterre 1995; Clemmens et al. 2005). During optimization, LQR control uses an objective function that needs to be minimized, which consists of all water deviations of all pools and all changes in control flows. These variables are squared to penalize both positive and negative deviations and changes. The objective function J may look like Eq. (6-20):

$$J = \sum_{k=0}^{\infty} \{P_{e1} e_1^2 + P_{e2} e_2^2 + \cdots + P_{\Delta Q1} \Delta Q_1^2 + P_{\Delta Q2} \Delta Q_2^2 + \cdots\} \tag{6-20}$$

where k is the time step index over an infinite horizon; P_{ei} is the penalty on the ith water level deviation from target level e_i; and $P_{\Delta Qj}$ is the penalty on the jth control flow change ΔQ_j.

Here, the objective function is written as a single equation for the sake of simplicity, although LQR is, in fact, defined in matrix form. Minimizing this objective function using appropriate software results in a feedback matrix that links all the states in the model to all control flow changes. Similar water level deviations in all pools will occur (and, therefore, no disturbance amplification) when penalties P_e are taken the same for all pools. Smoother control is achieved with larger values for the $P_{\Delta Q}$ penalties relative to the P_e penalties, whereas tighter control is achieved with smaller values for the $P_{\Delta Q}$ penalties relative to the P_e penalties.

6.5.7. *H*-Infinity Control

A controller that can be implemented as a SISO as well as a MIMO control system is H_∞ control. Instead of using a quadratic objective function that is minimized over a time horizon for controller design, the controller is designed in the frequency domain. In this way, the resulting controller functions properly for slow-moving situations, as well as for fast-moving situations, even on disturbances that excite the resonance waves of a canal pool. The specifications for the controller design are given in terms of required closed-loop performance and disturbance rejection. Extensive literature is available for this topic; see in particular Litrico and Fromion (2009).

6.5.8. Model Predictive Control

When limitations in structure flows are present, anticipation may be necessary. *Model predictive control* (MPC) is a structural methodology to deal with this. As with LQR, it uses an objective function with penalties on different, conflicting objectives. The prediction horizon for MPC is a certain finite length. After initialization from measurements of the prediction model with the present situation, an optimization is executed online to find the optimal structure flows over the horizon. After the optimization, the first control action is implemented, and after one control time step the entire procedure is repeated. In the optimization, limits are imposed on the control flows; thus, future problems and violations of constraints are avoided. Check the extensive literature on this topic for more information (e.g., Camacho and Bordons 1999; van Overloop 2006). The primary drawback of MPC is that an optimization needs to be run online on a powerful computer, and this may take several minutes before the optimal actions are calculated.

Often MPC is not necessary in irrigation canals, generally because the offtake flows in the schedules usually are preprocessed and optimized not to exceed the structure flow capacities anywhere in the canal (van Overloop 2006).

6.5.9. Proportional Integral Flow Control

The PI–algorithm is a general method and also can be used for controlling the flow at a target flow, Q_{ref}. This target can be constant or imposed by a higher hierarchical control layer (e.g., the PI–controller on the water level as described previously). In the latter case, the flow target is time variant. The flow measurement usually is taken downstream of the control structure, for example, using a measurement flume. This method is preferable over inverting the gate equation to what is referred to in Chapter 4 as feedforward flow control. The PI-controller for the flow is defined by Eqs. (6-21) and (6-22):

$$e_Q(t) = Q(t) - Q_{\mathrm{ref}}(t) \tag{6-21}$$

$$\Delta Q_c(t) = K_P(e_Q(t) - e_Q(t - t_s)) + K_I e_Q(t - t_s) \tag{6-22}$$

where Q is the measured flow; and e_Q is the deviation between measured and target flow.

The K_P and K_I can be taken as

$$K_P = -0.354,\ K_I = -1.175 \times \frac{t_c}{\max(\tau, t_c)} \tag{6-23}$$

where τ is the delay time between control flow and measured flow.

Field implementations often use only a proportional controller, especially when the structure and measurement location are close. In that case, the proportional gain, K_P, can be selected rather high, which results in good performance and only a small static deviation from the required flow.

6.6. COMBINING FEEDBACK AND FEEDFORWARD CONTROL

Often the offtake flows are known beforehand. Users have filed requests for water, and these orders are recorded. Now, if the delay times of all pools are known, water can be released in advance, and all structures can be scheduled to pass the appropriate amount of water at the right time. In this way, the flow is routed through the canal. Methods to compute those times were discussed in Section 6.4 of this chapter on Routing Demand Changes through a Canal. Note that combining feedback and feedforward control is only possible when both use incremental control of flow rates or changes from the existing flow.

Figures 6-1 and 6-2 represent the implementation of feedforward on top of feedback control for downstream control direction and upstream control direction, respectively, where Q_{FB} is the required feedback flow based on the deviation in the pool under control by that controller, Q_{DC} is the feedforward flow that decouples a pool from the feedback control actions of the adjacent pool, Q_{FR} is the flow imposed by the feedforward scheduler by which the flow is routed through the canal, and Q_C is the summed control flow of a structure. The most upstream structure is the headgate. Any offtake flows are left out of the figures. This means that, in this case, all feedforward routing flows are based on the amount and timing of the schedule of the most downstream structure. Obviously,

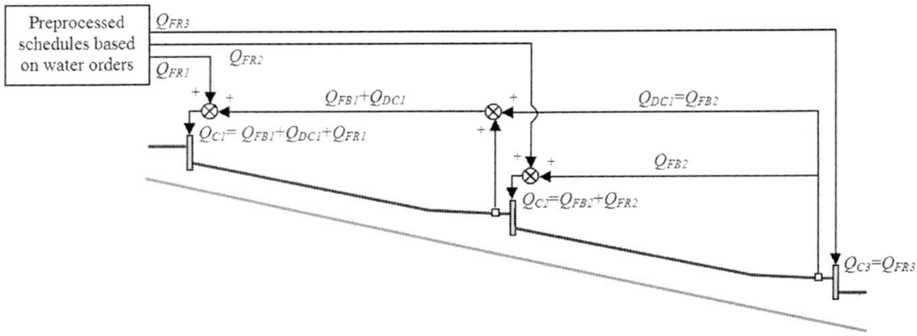

Fig. 6-1. Combined feedforward, feedback control, and decoupling for downstream control direction

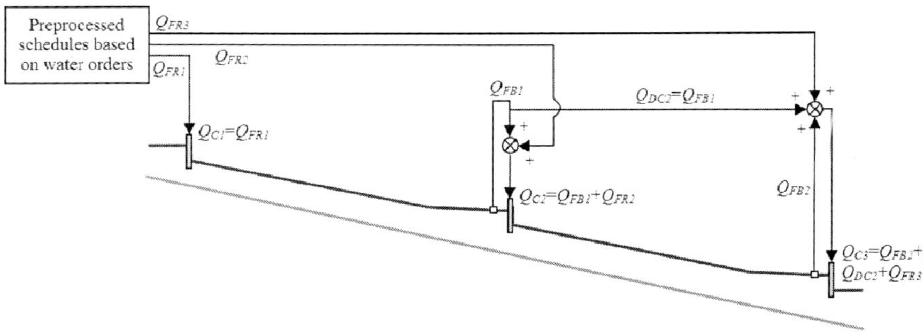

Fig. 6-2. Combined feedforward and feedback control for upstream control direction

adding offtake flows in the pools requires adding scheduled flows in the feedforward routing.

In a perfect world where everything could be exactly measured, the controlled flow could be set precisely and without wave dispersion, and feedforward control actions would lead to absolutely no deviations from the water level target. Thus, theoretically, feedforward is a very powerful control mechanism. However, the delay time never is known exactly and flow changes are dispersed as the wave moves downstream, so in actual circumstances feedforward can be rather inaccurate. For that reason, feedforward should always be implemented in conjunction with feedback control. For manual control, the operator provides the feedback actions.

6.7. REFERENCES

Bautista, E., and Clemmens, A. J. (2005). "Volume compensation method for routing irrigation canal demand changes." *J. Irrig. Drain. Eng.*, 131(6), 494–503.

Belaud, G., Litrico, X., and Clemmens, A. J. (2013). "Response time of a canal pool for scheduled water delivery." *J. Irrig. Drain. Eng.*, 139(4), 300–308.

Camacho, E. F., and Bordons, C. (1999). *Model predictive control*, Springer, New York.

Clemmens, A. J., Bautista, E., Wahlin, B. T., and Strand, R. J. (2005). "Simulation of automatic canal control systems." *J. Irrig. Drain. Eng.*, 131(4), 324–335.

Litrico, X., and Fromion, V. (2009). *Modeling and control of hydrosystems*, Springer, London.

Malaterre, P. O. (1995). "PILOTE, optimal control of irrigation canals." *Water resources engineering*, W. H. Espey, Jr., and P. G. Combs, eds., ASCE, New York, 338–342.

Schuurmans, J. (1997). "Control of water levels in open channels." Ph.D. dissertation. Delft University of Technology, Delft, The Netherlands.

Van de Vegte, J. (1990). *Feedback control systems*, Prentice Hall, Upper Saddle River, NJ.

van Overloop, P. J., Schuurmans, J., Brouwer, R., and Burt, C. M. (2005). "Multiple-model optimization of proportional integral controllers on canals." *J. Irrig. Drain. Eng.*, 131(2), 190–196.

van Overloop, P. J. (2006). "Model predictive control on open water systems." Ph.D. dissertation. Delft University of Technology, Delft, The Netherlands.

CHAPTER 7

VERIFICATION OF CONTROLLER PERFORMANCE

7.1. INTRODUCTION

The users of canal automation want to be sure that the automatic control system will function in a way that is useful for their operations. Two essential questions must be answered. First, are we controlling the right processes? Second, is the control functioning in an acceptable manner? This chapter discusses the process that the control system designers and integrators should go through so that they can document a successful automation implementation to users. Users should also be aware of these requirements so that they can specify the requirements for automation contracts. Chapter 6 discusses canal control algorithms. Such algorithms require the development of control parameters or coefficients. Once the algorithm and parameters are selected, they are, in combination, referred to as a "controller." Two different controllers for a canal can be based on the same algorithm but with different coefficients, or be based on different algorithms.

At the start of any automation process, the users and automation developers often have high expectations. Once the automation is in place, performance may not meet any of these expectations. Sometimes users do not fully understand the implications of decisions made regarding control strategies until the system has been implemented. At the same time, designers cannot always predict the performance of their control system design. With canals, automatic controls can only do so much. The actual level of control often is limited by the physical infrastructure and the nature of open channel flow. As the saying goes, "You can't push a rope." The time to identify the limitations is during the initial design of the system.

Consider automatic upstream control of water levels, which is a common first step in automation. After implementing upstream water level control, the canal operations authority may discover that the implemented "automation" does not provide for the correct flow rate at the customer's delivery gate. In other words, it does not provide total automation. If an upstream customer takes extra water, a downstream customer may be shorted. Control of the flow rate at the canal headgate is necessary to provide the correct amount of flow to the canal. Where canal spills are to be minimized, the canal operator still must change the canal headgate flow to service a customer's new order. And it still takes time for water to pass through the canal pools before it arrives at the user's gate. The operator still must monitor water deliveries and spills (feedback) and then adjust the headgate flow to balance demands. This control method can provide improved operations and control, but the user must understand what it does and does not do.

Experienced canal automation designers may be able to predict the level of performance possible for a given control scheme. Unfortunately, there have been a limited number of canals automated with a limited amount of automation. So to rely solely on experience is not prudent at the current time. Preimplementation performance tests are recommended to document the potential performance of the chosen canal automation strategy, typically through simulation modeling. Once the system designer communicates to the canal operators the anticipated performance criteria such that operators understand the limitations of canal automation for their system, the operators may choose to implement other features, such as small reservoirs, instead of or in addition to canal automation. This also assists in getting the operational personnel comfortable with the automation and understanding the benefits and limitations of the automation.

Good communication between the control system designer and the operations personnel is essential to avoid misunderstandings. A good way to avoid communication problems is for the operators to be able to describe how the control system is to perform with specific tests that should be used to evaluate control system performance.

7.2. PERFORMANCE TESTING ISSUES

This section deals with the testing of automated canal structure logic. There are two categories of automation sites for irrigation canals, listed here with examples. The performance testing requirements for these two types of automation sites are quite different.

1. Simple, independent, devices that perform
   a. Automatic flow control of a canal headgate;
   b. Upstream water level control at a single gate, for example, a canal spill;

   c. Control of the flow into or out of a reservoir; and
   d. Variable speed control of a single pump.
2. Complex configuration and/or multiple devices that perform
   a. Upstream control at a series of check gates and
   b. Downstream control of the level at the downstream end of a canal pool.

"Independent" implies that the parameter being controlled is not influenced by other automatic devices. The simple devices imply that the controlling device (e.g., gate) is physically very close to the measured variable (the level or flow that is being controlled) for canals within, for instance, 100 m. An automated device is considered "complex" if it is influenced by other controlled devices or if there is a significant travel time for flow changes to reach the measurement site, particularly where information from one field unit (remote terminal unit [RTU] or programmable logic controller [PLC]) must communicate with another.

For simple devices, automatic controls can be installed, tuned, and tested in the field. For complex devices, performance testing with unsteady-flow simulation models is recommended. This provides end users with a better understanding of the capabilities of the canal automation system being proposed. If users are unfamiliar with automatic devices, they may want even simple devices to be demonstrated with a simulation model. This should be the user's prerogative. Users with extensive experience with complex devices may forego simulation testing, but this is rare. The unsteady-flow model also can be used effectively to educate the staff members of the operating agencies on the use of the control system and allow the operators to investigate how to deal with various conditions, including emergencies.

As described in Chapter 6, the intent of automatic devices is to move the value of a controlled variable (usually water level or flow rate) toward a desired value. Typically the field unit obtains a reading from a sensor and computes the value of the controlled variable. For example, the field unit reads a value from a pressure transducer and computes the water depth. The automated device makes corrections (e.g., changes in gate position) until the computed water depth equals the desired water depth. However, any condition that causes an incorrect sensor reading causes the controller to move the water level to the wrong position. This is not the fault of the automatic control algorithm. The same issue applies to flow rate control. If the flow measurement device is inaccurate, the automatic control moves the flow rate to the wrong value. Errors in the flow measurement device calibration cannot be overcome by the automatic control device. Separate specifications are needed to calibrate water-level transducers and flow measurement devices. Control system users must also be prepared to dedicate resources to the maintenance and calibration of the control system components that quite often require the employment of staff with a different skill set than needed for manual operations.

Most automatic control devices adjust a gate or valve until the desired water level or flow rate is obtained. Most gate or valve position sensors are digital, which implies a minimum resolution. For example, one increment in the sensor output might represent 10 mm in the physical gate position. Thus, the gate position sensor knows the gate position only to within ±5 mm, and the control system would only be able to control the gate within ±5 mm. This creates a random error in the gate control. If the gate flow is sensitive to gate position changes, it may not be possible for the control system to set the correct flow rate or to maintain the correct water level. For a water level controller, this can cause the water levels to oscillate, because it cannot move the gate to the "perfect" position to balance inflow and outflow to the canal pool. If oscillations occur, it is important to distinguish between an inadequate control algorithm, improper controller tuning, and poor resolution in some physical system components.

7.3. PERFORMANCE TESTING WITH UNSTEADY-FLOW SIMULATION MODELS

Simulation models are a useful tool for determining the potential performance of canal automation. Many canal automation developers routinely test control algorithm performance for even a single gate with unsteady-flow simulation. This allows the gate to be tested under a variety of flow and operating conditions. Without the use of unsteady simulation, it would take a long time to experience this wide range of conditions in practice. An automated gate may work fine in the normal operating range and suddenly fail at high or low flow (not normal) conditions. Simulation modeling can help identify and avoid this kind of problem.

Modeling of irrigation canal pools is discussed in Chapter 5, where simplified models of water level response are used for control algorithm design (Chapter 6). Here, we want to simulate water flow in the canal with a complete solution of the unsteady-flow (Saint Venant) equations. Numerous simulation models are available for solving these equations. However, few are set up specifically for canal automation, so some care must be taken to ensure that they are implemented appropriately. The following are some issues that should be addressed:

1. For multiple pool canals, the model should be fully implicit and be able to model all pools.
2. A time weighting coefficient of roughly 0.55 to 0.7 should be used. (Many programs use a default value of 1.0, which should be changed before running simulations, since this tends to dampen wave action.)

3. The time step of simulation should be consistent with the canal dynamics. A sensitivity analysis of time step can be used to determine an appropriate time step (i.e., try smaller time steps until results no longer change). Typical canals can start with 1 to 5 min (Clemmens et al. 2005). Smaller time intervals are needed in some cases.
4. The simulation model must be capable of modeling any gates, weirs, and pumps that are part of the control system, including structures in parallel. If structure calibrations are not accurately known, errors in calibration should be included in the simulation of controller performance (see Section 7.6, Test Case Examples).
5. The automatic control routines must be separate and independent from the simulation model itself, so that there is real separation between the hydraulic calculations and the control calculations.
6. The simulation model must include restrictions on gate movement that match the gate resolution of the physical system.
7. Models should be calibrated against physical observations, at least for steady-state conditions at high and low flows.
8. Model results should be presented to users in an understandable format.

The better the simulation model represents the actual physical conditions of the canal and structures, the more similar simulation results are to field observations. For simple devices, simple step tests are sufficient to evaluate control accuracy and stability (see Chapter 5). As discussed, these tests do not require simulation modeling. These can be conducted with field tests (see Chapter 8).

For complex devices, performance of the control system can include a variety of situations. These might include

- Routine operations at high flow;
- Routine operations at low flow;
- Off-season operations (e.g., when operating conditions or procedures differ);
- Routine operations at a time of year with different canal roughness; or
- Extreme conditions, such as emergency shutoff by customer or gate.

Some of these tests can be made only through simulations, whereas others may require field tests.

It is important that the automation components function properly during routine operations and that the operators have been properly oriented on the control systems. These simulation tests ensure that the control system designer fully understands real operating conditions. This may also identify limitations in the automation and help define what the automation can and cannot do. In these tests, it is also important to consider measurement error. In a simulation model, gate settings and discharge predictions can be exact, whereas field information never provides the exact water level, gate position, or flow rate. This should be considered carefully when developing test scenarios.

Canal automation algorithms generally are tuned for a specific set of conditions. Changes in flow resistance caused by weeds, grass, and sediment can slow down the responsiveness of the canal. If the controller does not consider these conditions, performance will degrade. Operators need to be informed of the nature of these performance degradations and how to compensate for them.

Modernization efforts that change the canal prism, such as lining an earth canal, also can alter the canal's responsiveness. Automatic gates tuned for an earthen canal may perform differently once the canal is lined. This can be tested with simulation modeling.

7.4. PERFORMANCE MEASURES

Commonly, canal automation is intended to accomplish one or more of the following:

- Supply water to users at some desired (perhaps variable and unknown) rate of flow over a specified time interval;
- Control water levels better;
- Reduce operator efforts;
- Reduce spills;
- Conserve water; and
- Perform operations that generally are impractical if done manually (e.g., downstream control).

For sites with manual offtake (turnout) structures, a constant water level implies a constant discharge. Thus, many automatic canal control systems work to maintain constant water levels. In other cases, control of flow rate is accomplished directly. Therefore, there are two control performance objectives for automatic control of devices or systems: water level control and discharge control. The evaluation of performance for these two types of control may differ.

7.4.1. Water Level Control

The ASCE Task Committee on Canal Automation Algorithms (Clemmens et al. 1998) developed a set of performance measures for a series of downstream control test cases. These performance measures were to be specified for each canal pool and included

- Maximum water level errors;
- Average water level errors;
- Steady-state errors; and
- Changes in gate movement or discharge.

For water level control, the change in gate movement is included in the performance criteria, because an overly aggressive controller (for example, an aggressive tuning for a given control algorithm) changes the gate position too often and too rapidly. Wear and tear on the gate motors, seals, and other parts is an issue. But the main reason for limiting gate movement is to ensure stability and to avoid oscillations or water level cycling. Thus, evaluation of water level control performance needs to consider both water level deviations and gate position changes. With so many indicators, it can be hard to judge the merits of one control algorithm (and its tuning) over another, unless one controller is superior in all indicators. And although many control routines work well in simulation models, they may not always work well in the field because of the field programming, PLC limitations, and equipment selection details. However, if control looks inadequate with the simulation model, excellent field equipment and programming cannot overcome an inferior controller design.

The task committee also discussed the use of quadratic indicators, although they were not included in their comparisons. Quadratic indicators for water level essentially integrate the first two criteria as listed, where the sum of the square of the water level deviation is the indicator of performance. The ASCE test cases went from one steady state (i.e., a set of flow rates and deliveries along the canal) to another. The test case criteria determined how many extra flow rate (or gate position) changes were made, in addition to the changes actually needed to change states. Such an indicator can be used for comparing controllers for specific test cases but is not a general indicator of true controller performance.

Quadratic indicators also can be used for gate flow or position changes, where the performance indicator is the sum of the square of the gate flow changes. Many algorithms for canal control attempt to minimize quadratic errors for both water levels and gate flow changes simultaneously. Because these have different units and have a different effect, weighting coefficients are needed to arrive at a single performance value for design. Selection of the appropriate weighting coefficients between water level deviations and flow rate changes is also somewhat subjective. These quadratic indicators for water level and discharge only indirectly consider the following:

- *Disturbance amplification* occurs when the control algorithm at individual check structures in series along a canal are tuned individually and not for the canal as a whole. For upstream control, disturbances grow in the downstream direction. A minor disturbance at the first gate becomes a major disturbance at the last gate.

- *Robustness* is an indication that the controller performs well over a wide variety of conditions.
- A system is *unstable* if a disturbance causes oscillations that grow without bounds. Use of a water level controller can cause water levels to oscillate, even if the system is stable. If the oscillations are large enough, the controller is not acceptable, even if the control is stable.
- Small *oscillations* can be caused by the inability of the gate controller to make small gate movements. They also can be caused by a water level deadband. (A minimum gate position change rather than a water level deadband is recommended.)
- *Steady-state errors* are errors that are not corrected, even after a long time. They can be caused by a weak controller. For a proportional integral controller, a small proportional and integral gain will result in a controller that moves the water level toward the set point very slowly.
- Small controller gains may be required to avoid oscillations and instability. So there is sometimes a trade-off between these two conflicting requirements: fast response and stable response. In general, slow response and a small steady-state error are preferred over large oscillations and instability. The intent of the control designer is to find an appropriate balance. Chapter 6 discusses control design procedures that help to find this balance.

7.4.2. Discharge Control

Control of discharge at a canal structure usually deals with a single flow measurement and a single gate that is used to change the flow. Thus, it tends to be less complicated than water level control. Fundamentally, the issues are similar:

- How long does it take to change to the new flow rate?
- Does the flow oscillate?
- Is there a steady-state error?
- Is the number of gate adjustments used to adjust the flow excessive?
- Is the control stable?
- How reliable (consistent) is the discharge measurement?
- How accurate is the discharge measurement?
- Does control of other structures cause difficulty with maintaining a constant flow?

Performance measures similar to the water level control performance measures can be used, but note that they consider flow rate, rather than water level. Again, there is a trade-off between stability (i.e., through minimizing the size and number of gate adjustments) and fast response. Automatic control of flow typically is much faster than control of water levels.

7.5. ASCE TASK COMMITTEE TEST CASES

At the time that the ASCE test cases were developed, distance downstream water level control algorithms were under their initial phase of development. Those control methods are now well developed theoretically, but experience has shown that distant downstream controllers have somewhat limited performance. Test Case I is a good example where unknown disturbances downstream can cause extreme changes in water levels and can limit flow to other users. The results from these tests clearly show that feedforward control or routing of known flow changes is essential for effective control of such canals. Changing turnout flows downstream on demand is just not feasible if the canal storage is small and the time for water to travel is long (i.e., relative to the time for a flow change to dewater the canal).

The performance parameters for the ASCE test cases from Clemmens et al. (1998) are described here.

Maximum Absolute Error (MAE)

$$\mathrm{MAE} = \frac{\max\left(\left|y_t - y_{\text{target}}\right|\right)}{y_{\text{target}}} \tag{7-1}$$

where y_t is the water level at time t; and y_{target} is the target water level.

Integral of Absolute Magnitude of Error (IAE)

$$\mathrm{IAE} = \frac{1}{T}\int_0^T \frac{\left|y_t - y_{\text{target}}\right|}{y_{\text{target}}}\,dt \tag{7-2}$$

where T is the time period of the test.

Steady-State Error (StE) In Clemmens et al. (1998), StE was defined as MAE during the last period of the test, supposedly after the main control actions had taken effect. Using the average absolute error during this period probably would have been a better choice (i.e., IAE). Note that MAE, IAE, and StE are applied to each canal pool separately, and they have no units.

Integrated Square of Error (ISE)

$$\text{ISE} = \sum_{i=1}^{N} S_i \frac{1}{T} \int_0^T \left| y_t - y_{target} \right|^2 dt \tag{7-3}$$

where S_i is the relative penalty for water level errors in pool I; and N is the number of pools. Typically, these penalties are equal for all pools. ISE is somewhat of a trade-off between large maximum errors (MAE) and large average absolute errors (IAE). It also is summed over all pools, rather than having to evaluate each parameter for each pool. ISE is shown here in dimensioned form.

Integrated Absolute Gate Movement (IAW)

$$\text{IAW} = \sum_{t=\Delta t}^{T} \left| w_t - w_{t-1} \right| \tag{7-4}$$

where w_t is the check gate position at time t; and Δt is the time step between possible gate movements. Note that t ranges from Δt to T and $t - 1$ ranges from 0 to $T - \Delta t$.

Integrated Absolute Discharge Change (IAQ)

$$\text{IAQ} = \sum_{t=\Delta t}^{T} \left| Q_t - Q_{t-1} \right| \tag{7-5}$$

where Q_t is the check gate discharge at time t. Note that IAW and IAQ have units. They are not dimensionless. Because most canal control design methods related gate discharge change to water level errors, IAQ is used to evaluate performance rather than IAW. Most controls prescribe changes in gate discharge, which are converted to gate position change through the head-discharge relationship of the gate. For structures with multiple gates and weirs, this provides more straightforward control.

In the test cases where changes were made in the flow delivered to users, check gate flows went from one steady state to another. Because this change was expected, this amount of change in either gate position or discharge was subtracted from the equations as stated previously. This did not change the comparison between different controllers, as the same number was subtracted in all cases. However, when trying to examine the trade-offs between the water level deviations versus the gate flow changes, it should be considered.

Integrated Square of Discharge Change (ISQ)

$$\text{ISQ} = \sum_{i=1}^{N} R_i \sum_{t=\Delta t}^{T} |Q_t - Q_{t-1}|^2 \tag{7-6}$$

where R_i is the relative penalty for gate flow changes in pool i. Clemmens and Schuurmans (2004) chose to adjust R_i in proportion to capacity. ISQ also is not dimensionless.

Linear Quadratic Regulator (LQR)

$$\text{LQR} = \text{ISE} + \text{ISQ} \tag{7-7}$$

where ISE is in dimensioned form (i.e., no y_{target} in denominator). A matrix expression for LQR is provided in Chapter 6. Note that use of Eq. (7-7) requires values for S_i and R_i. This selection is a major factor in controller tuning. Essentially it provides the trade-off between rapid response and stability.

7.6. TEST CASE EXAMPLES

A number of different controllers were tested. Here the results from Clemmens and Wahlin (2004) for Test Case 1-1 are presented because they show examples of the control issues of interest in this chapter. This does not imply that the control methods in this manual are preferred over other control methods. Here, the results are presented for two different controllers. Both controllers were tuned through optimization with state-space control as a linear quadratic regulator (LQR). Both were tuned with the same tuning factors. In one case a series of single-input, single-output (SISO) proportional integral (PI) controllers was designed. In the second case, a single multi-input, multi-output (MIMO) controller was designed, referred to as an *optimal controller*. For feedforward control, the volume-compensation method of Bautista and Clemmens (2005) was used, as discussed in Chapter 6. Simulations were performed in SOBEK with a 1-min time step.

ASCE Test Canal 1 is a relatively steep canal with eight pools with little storage. Test Case 1-1 has a change in flow rate at several turnouts at 2 h and at 14 h. At 2 h, offtakes 3 and 4 both increase their delivery flow by 0.1 m^3/s. The canal goes from 0.8 to 1.0 m^3/s. These changes are known to occur by the controller. At 14 h, offtakes 4 and 5 decrease their flow by 0.1 m^3/s. The canal flow thus goes from 1.0 to 0.8 m^3/s. These offtake changes are not known by the controller, even after they occur. The controller only can observe changes in water levels.

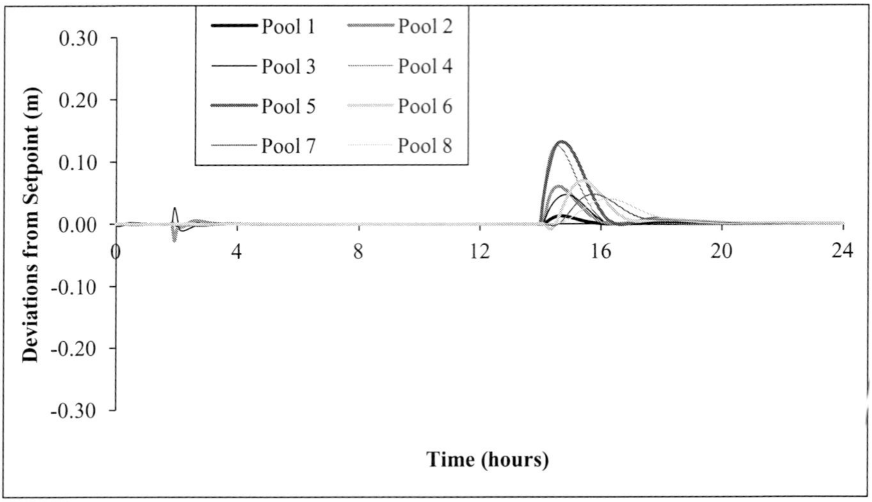

Fig. 7-1. Simulation results for Test Case 1-1, optimal controller, no gate restrictions, tuned conditions
Source: Fig. 3 in Clemmens and Wahlin (2004); reproduced with permission

Fig. 7-1 shows the resulting water levels when the optimal controller was used. At 2 h, one can see very little variation in the water levels. This means that the timing from the feedforward control was reasonably good. At 14 h, large deviations occur on the order of 12 cm. Deviations are worst in pools 4 and 5 where the changes occurred, but these changes propagated to other canal pools, both upstream and downstream. It took about 3 h for the changes to be reduced to acceptable levels and the oscillations to dampen out. A controller with different penalty weighting perhaps could reduce the magnitude of these changes by responding sooner, but as is shown in the next example, this may not be useful. This is essentially as good as a downstream controller can do in controlling this disturbance for this canal under ideal conditions. If this is not acceptable, this method should not be applied.

The ASCE test cases included a requirement that the controllers be tested under "untuned" conditions. In this case, the Manning's *n* was changed from 0.014 to 0.018 and check gate calibrations in the simulation had to differ from that assumed in the control calculations by 10%, and offtake flow changes had to be 5% higher. Because canal conditions change and flow rates often are uncertain, this was considered to be a more realistic test. Fig. 7-2 shows the results of the optimal controller for the untuned conditions. It can be seen that controller performance has degraded. More deviation at 2 h suggests that the routed flow changes did not match. This

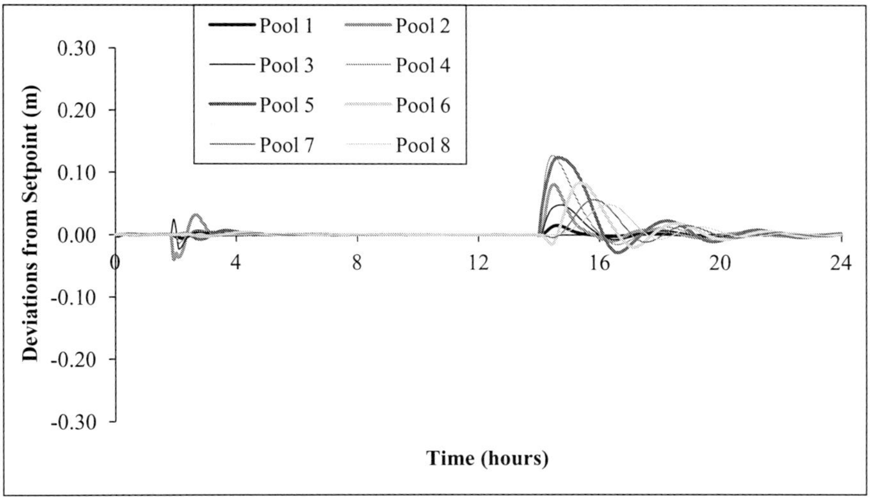

Fig. 7-2. Simulation results for Test Case 1-1, optimal controller, no gate restrictions, untuned conditions
Source: Fig. 5 in Clemmens and Wahlin (2004); reproduced with permission

occurred both because the outlet flow and check gate flow no longer match and because the flow arrived later due to the increased canal roughness. At 14 h, the deviations are slightly more and have a longer duration. Oscillations still can be seen as much as 6 h later.

Fig. 7-3 shows the results for the SISO PI controllers for untuned conditions. For tuned conditions, the results were not much different from the optimal controller (not shown). However, for untuned conditions, deviations are significantly greater, and significant oscillations occur. This kind of oscillatory behavior is not desirable. Observations of this kind of behavior suggest that such a controller should not be implemented on this canal.

Comparisons of Figs. 7-2 and 7-3 demonstrate the need for MIMO control for downstream control in some cases. These controllers were tuned with the same conditions and same performance criteria. This test case canal was based on the WM canal in Maricopa, Arizona. Clemmens and Strand (2010) tested these controllers on the actual canal and reached the same conclusion: SISO PI controllers should not be used on this canal.

The ASCE test cases limited the minimum amount of gate movement to 0.5% of gate height. Limits in the amount a gate can be moved can have a significant influence on controller performance. Some automation efforts have failed because they did not recognize the significance of this restriction. Fig. 7-4 shows the result of the optimal controller under tuned

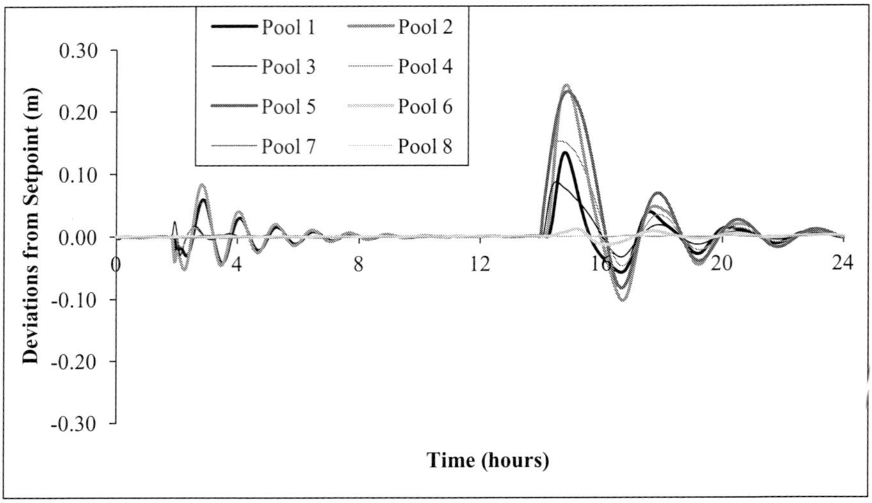

Fig. 7-3. Simulation results for Test Case 1-1, series of simple PI controllers, no gate restrictions, untuned conditions
Source: Fig. 6 in Clemmens and Wahlin (2004); reproduced with permission

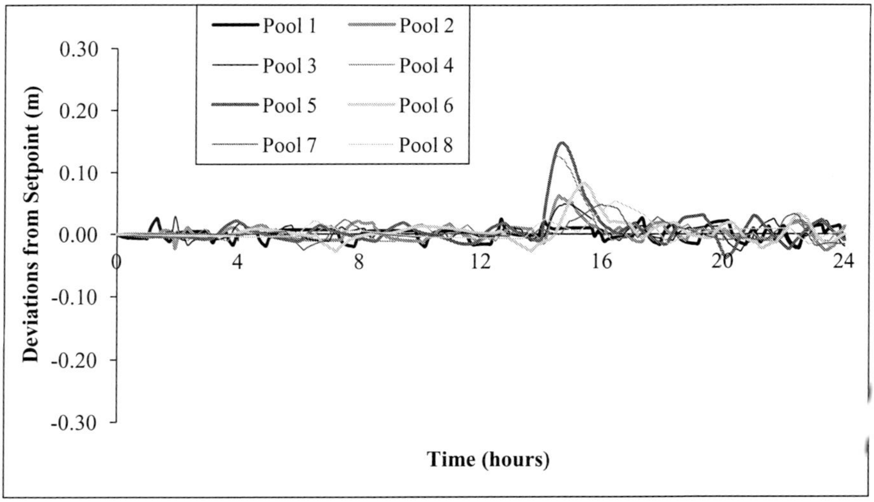

Fig. 7-4. Simulation results for Test Case 1-1, optimal controller, gate restrictions (0.5%), tuned conditions
Source: Fig. 9 in Clemmens and Wahlin (2004); reproduced with permission

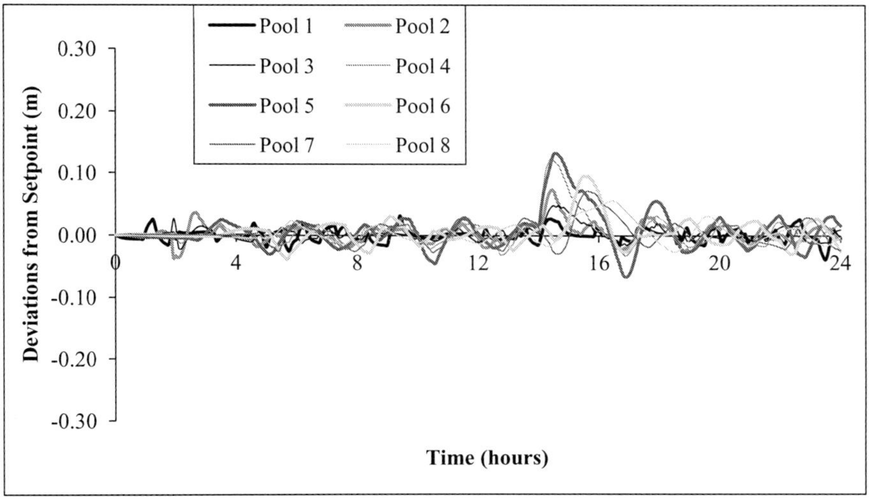

Fig. 7-5. Simulation results for Test Case 1-1, optimal controller, gate restrictions (0.5%), untuned conditions
Source: Fig. 12 in Clemmens and Wahlin (2004); reproduced with permission

conditions with restrictions in gate movement. This should be compared to Fig. 7-1. Although the deviations caused by the unscheduled changes at 14h are not much higher, these controllers cannot seem to stabilize the water levels. With gate movement restrictions, the gates simply tend to oscillate because they cannot move to the correct position. For these tests, the canal depths were 1.9 to 2.5m, whereas 0.5% of 2m is 1cm. Current technology can result in gate movements as low as 1mm. Thus, this restriction for the ACSE test cases was a bit too restrictive, but it should be a red flag to those implementing automation: understand the precision to which the gate position is controlled.

Fig. 7-5 includes both gate restrictions and untuned conditions. Interestingly, the untuned conditions do not seem to make a lot of difference. At this level, the gate restriction is more of a concern. Fig. 7-6 represents the SISO PI controllers with restriction on gate movement and untuned conditions. This shows that the oscillatory behavior does not dampen out and further reinforces the conclusion not to use SISO PI controllers on this canal.

7.7. ADDITIONAL CONSIDERATIONS

With upstream control, the most significant issue is disturbance amplification. If multiple gates are in series, poor control tuning can result in

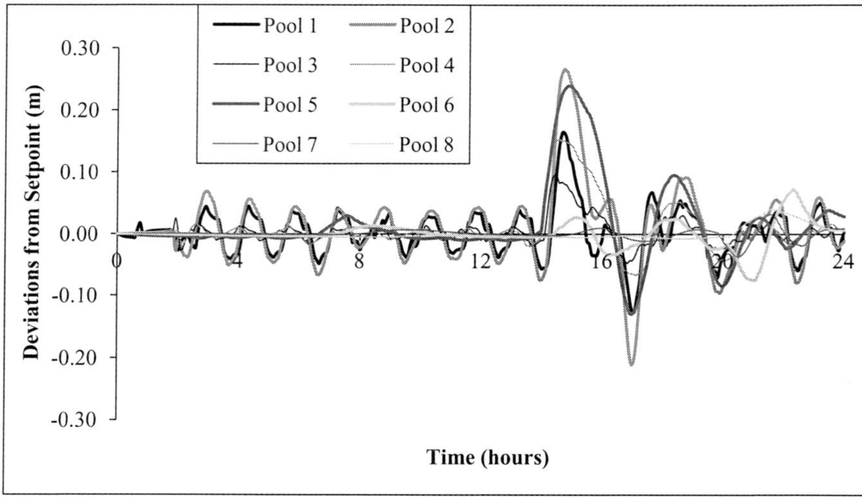

Fig. 7-6. Simulation results for Test Case 1-1, series of simple PI controllers, gate restrictions (0.5%), untuned conditions
Source: Fig. 13 in Clemmens and Wahlin (2004); reproduced with permission

oscillations that increase in the downstream direction. Testing with simulation under untuned conditions should be used to determine if this could occur.

Many of the historic methods for control of water levels used a water level deadband to avoid oscillation in gate position and avoid the behavior shown in Fig. 7-4. However, this is not recommended. Rather, it is recommended to compute the control action and then restrict the gate movement to avoid oscillations. Using a water level deadband actually interferes with the control action as the water level passes through the set point.

The tests developed to determine if a controller will perform adequately should be reviewed with operating personnel, who often have useful insight into what conditions could cause problems with the control.

7.8. REFERENCES

Bautista, E., and Clemmens, A. J. (2005). "Volume compensation method for routing irrigation canal demand changes." *J. Irr. Drain. Eng.*, 131(6), 494–503.

Clemmens, A. J., and Schuurmans, J. (2004). "Simple optimal downstream feedback canal controllers: Theory." *J. Irr. Drain. Eng.*, 130(1), 26–34.

Clemmens, A. J., and Strand, R. J. (2010). "Downstream-water-level control test results on the WM lateral canal." *J. Irr. Drain. Eng.*, 136(7), 460–469.

Clemmens, A. J., and Wahlin, B. T. (2004). "Simple optimal downstream feedback canal controllers: ASCE test case results." *J. Irr. Drain. Eng.*, 130(1), 35–46.

Clemmens, A. J., Bautista, E., Wahlin, B. T., and Strand, R. J. (2005). "Simulation of automatic canal control systems." *J. Irr. Drain. Eng.*, 131(4), 324–335.

Clemmens, A. J., Kacerek, T. F., Grawitz, B., and Schuurmans, W. (1998). "Test cases for canal control algorithms." *J. Irr. Drain. Eng.*, 124(1), 23–30.

CHAPTER 8
IMPLEMENTATION OF CONTROL SYSTEMS

8.1. INTRODUCTION

This chapter addresses the practical implementation of canal automation as a project within a wider modernization program. The automation of an entire lateral or whole canal system is described, rather than just a pool or a few pools, because the challenges and benefits are much larger (Fig. 8-1). This chapter describes the sequence of required tasks, including the customization of control software; installation and commissioning of the field devices and central server software; progressive activation of control throughout the canal system; measurement of overall performance of the new system; and training of operations staff. Ongoing operational tasks and system maintenance also are discussed.

8.2. PROJECT INITIATION

The first step in any automation project is to determine the project objectives. These might include improvement of irrigator service levels, water savings, and better control of water levels; operational cost savings; or rate of return on investment. Ideally, the objectives can be quantified. Clear, quantified objectives help ensure success by aligning the irrigation district and implementation team and allowing for objective measurement of performance. More information can be found in Chapter 1.

Once objectives are clear, implementation proceeds along three broad streams of effort, as shown in Fig. 8-2: selection, installation, configuration, and testing of field devices; selection, installation, configuration, and

Fig. 8-1. An entire automated canal system in Australia showing automated checks, turnouts (left foreground), and communications tower (background) Source: Photo courtesy of Austrade, Sydney, Australia; reproduced with permission

testing of the IT infrastructure; and definition, design, rollout, and performance assessment of the control system and its operation. There is process flow across these three streams, and the interflow can vary from project to project. Cutting across these three streams is the need for training.

Some of the steps (shown in gray in the diagram) are the subject of other chapters in this manual. They include the selection of

- *Control objectives* such as the desired operating strategy (supply or demand orientation), level, volume, or flow control;
- *Instrumentation or field equipment* such as regulating gates, sensors, actuators, power supply, access to the gates, RTU/PLCs, HMI electronics for the RTU/PLC, and the software within the field equipment/RTU;
- *IT infrastructure or context* such as server and storage, virtualization, networking (WAN, LAN, etc.), enterprise management system, infrastructure software, and others;

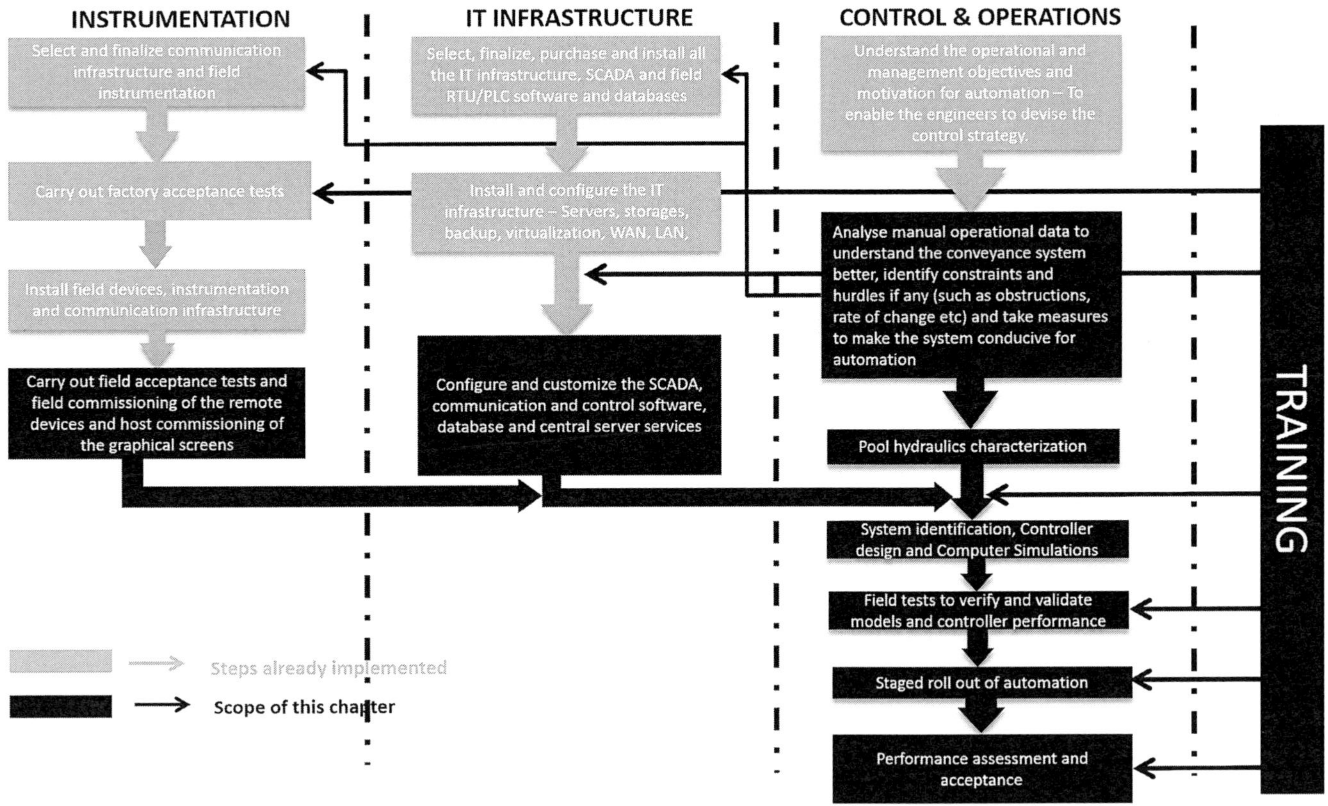

Fig. 8-2. Incremental process toward implementing control; note that there is process flow across these three streams and the interflow can vary from project to project

- *Software,* including SCADA software and associated databases; network visualization; and potentially for collecting water orders, managing demand, and planning deliveries; and
- *Communications infrastructure and protocols* such as optical fiber cable, VSAT, radio, and protocols (e.g., TCP/IP, DNP3, MDLC/IP, and MODBUS/IP).

A number of other generic tasks under project delivery and management such as the procurement and installation of field devices, communications infrastructure, and IT infrastructure are not included in this chapter, because they are well understood and published in other sources, such as publications of the Project Management Institute (PMI 2008).

8.3. CONFIGURATION AND CUSTOMIZATION OF SOFTWARE

The automation of an entire lateral or whole canal system usually requires the specification of two broad types of control objectives: site objectives and networkwide objectives.

- *Site management objectives* are concerned specifically with local goals such as maintaining a set flow through the site or maintaining immediate upstream or downstream water levels at set points. The algorithm to communicate with the sensors and actuators and the control algorithm generally reside in the RTU/PLC at the control gate. To enable the irrigation authority to monitor and operate the system remotely, only SCADA software and communications infrastructure are required.
- *Network control objectives* require that a site participate in control with its neighboring sites to meet systemwide or networkwide objectives (as well as local objectives). In its fullest configuration, the irrigators interact with the network and place orders. The orders are scheduled automatically to ensure that the canal is operated within its safe operational limits, and the information about irrigator orders is used by the control system. This capability requires completion of most of the infrastructure and associated services mentioned earlier, as is illustrated in Fig. 8-3.

Therefore, the required control software depends on the desired control objective. Control software can fall into several categories:

- *RTU/PLC* software resides in the RTU/PLC at the control gate to scan the sensors and drive the actuators. When integrated with SCADA, this software also should communicate with the central

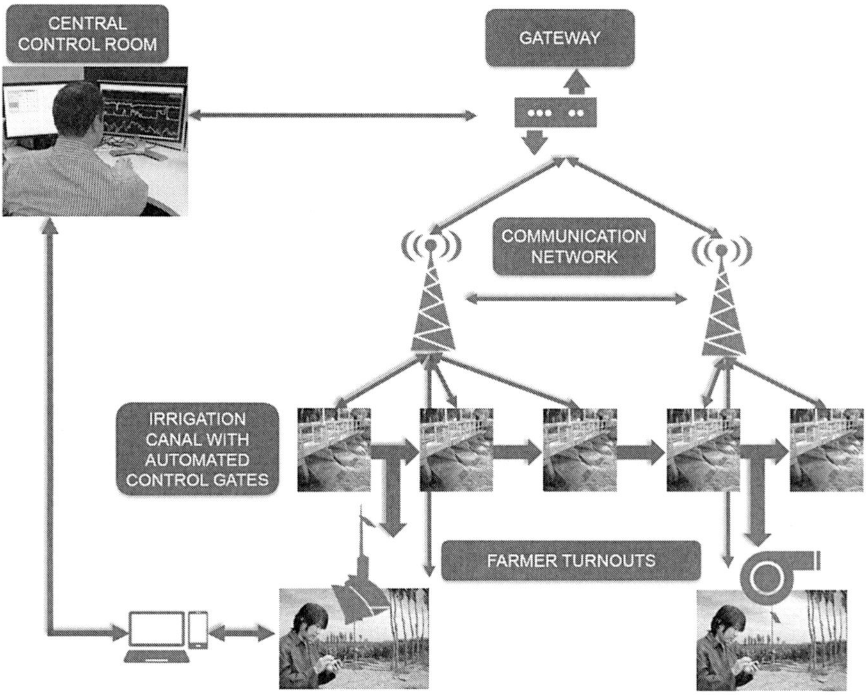

Fig. 8-3. Automation infrastructure and stakeholders

server. Depending on the control implementation and communications architecture, advanced control tasks such as flow control, immediate upstream and downstream control, distant downstream control, or peer-to-peer communication can be performed by the RTU/PLC software.

- *Communications* software resides in the communication hardware in the RTU/PLC at several locations that must communicate with each other: control gates, communication nodes, and the central server.
- *Supervisory control and data acquisition (SCADA)* software enables the canal operator to monitor and operate a control gate, a set of gates, or an entire irrigation network remotely.
- *Water ordering and demand management* software gives irrigators the ability to view and place orders and also ensures that orders do not exceed capacity and safe operational constraints.
- *Network representation* software provides the topology and interconnectivity information about the network to the network control and water ordering and demand management software (see Chapter 3).

Each type of software and its functionality is discussed briefly next, followed by the software configuration tasks that are required as part of implementation. A detailed explanation of SCADA software can be found in Chapter 3.

8.3.1. Remote Terminal Unit (RTU)/Programmable Logical Controller (PLC) Software

RTUs and PLCs can be considered the "brain" of the control gate. Both RTUs and PLCs are microprocessor-controlled electronic devices that provide control and monitoring functionality to a control gate.

Originally designed to replace relay logic, PLCs acquire analog or digital data through input modules and execute a program loop while scanning the inputs and taking actions based on these inputs. PLCs perform well in sequential logic control applications but suffer from overly specialized design, which results in limited CPU performance, inadequate communication flexibility, and lack of easy scalability when it comes to adding future requirements other than input–output (I/O) (Motorola 2007).

RTU design focuses on remote monitoring with control but with a higher demand for application communications and protocol flexibility. RTU designs tend to have greater CPU processing power, more programming flexibility, and broader communication support than PLC systems (Motorola 2007) and, hence, are used widely in open canal automation, where the sites are remotely located and require local processing power for lesser reliance on a central server. However, features of RTUs and PLCs are increasingly beginning to overlap, and many vendors sell RTUs with PLC-like features and vice versa.

RTUs often use their own proprietary programming platforms, whereas the IEC 61131 programming tools are popular with PLCs. Whatever the RTU/PLC, it should be programmed and configured to carry out the following functions:

- Scan the analog or digital inputs to obtain measurements from sensors for local control calculations or for transmission to the central server.
- Provide mapping interface for the data received from the instrumentation interface to the RTU database.
- Provide filtering and averaging of the scanned inputs to suppress unwanted noise.
- Communicate with the central server and neighboring sites, eliminating the need for a central communication hub to transmit information between sites.

- Provide stand-alone control such as flow, position control, or immediate upstream and downstream control.
- Provide algorithms for network control such as distant downstream level control or volume control.
- Calculate flow at the site, either through ratings tables, mathematical relationships, or other means.
- Ensure bumpless transfer when switching between control modes.
- Trigger alarms when process variables breach safe operating limits.
- Provide an interface to nonvolatile memory, allowing configuration parameters to be stored and ensuring that their values are retained after a power loss or reboot.

Some commercially available RTU/PLC programs such as Rubicon NeuroFlo are designed specifically for canal operations. This software can be purchased off the shelf and can be configured as per site-specific scenarios.

8.3.2. Communications Software

Communication to remote sites is essential for remote monitoring and control of field devices. Communication to the central server and between neighboring sites is possible via various mediums such as optical fiber cable, VSAT, or radio, as long as the protocol is designed to function in the selected medium. Communications software resides in the communications hardware in the RTU/PLC at the control gates, in the communications nodes, and at the central server. Communications hardware usually arrives preprogrammed and tested by the manufacturer.

Hardware redundancy is vital at all levels of the communications system to minimize the effect of a communication outage. The architecture of the communication system plays a crucial role in successful automation, as timely exchange of data between remote sites, the central server, and various services within the central server is vital to smooth operations. It also is crucial that all the sites in the network have their time synchronized and security enabled.

Usually the only configuration requirements for communications software involve specifying the frequency, bandwidth, correct link, baud rate, and a unique site address. The configuration can be carried out by connecting the communications hardware to a computer and using the appropriate software. Communications can be configured at the factory before being installed on site.

8.3.3. Supervisory Control and Data Acquisition (SCADA) Software

SCADA software resides in a central server and is used by the operators in a control center to monitor and operate a canal network in real time.

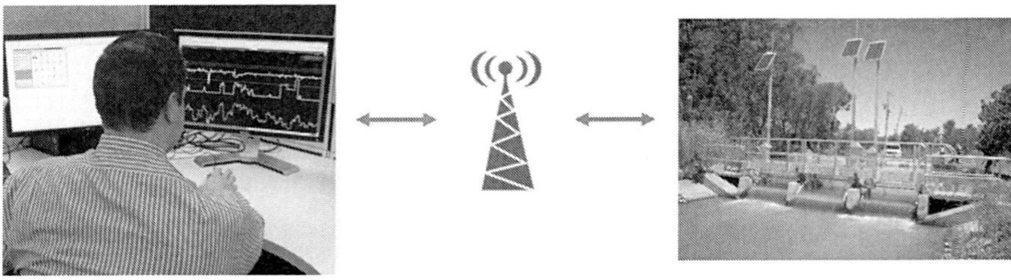

Fig. 8-4. An operator monitoring and controlling field device

SCADA software requires the RTU/PLC, communication, and IT infrastructure to perform their tasks. Operators use SCADA software to monitor flows, water levels, gate positions, the status of components, and any other variables considered relevant to be displayed on SCADA screens. The operators also can open or close control gates remotely and operate other field devices such as pumps.

The information received by the SCADA software is stored in a database and is made available in real time to connected computers. Many SCADA software packages are commercially available, including Rubicon SCADAConnect (Fig. 8-4), CITECT, ClearSCADA, Emerson Open Enterprise, ABB SCADA, GE-HMI/SCADA iFix, Reliance SCADA , IGSS, and others. With these packages, a working SCADA system often can be built without needing to write any program code. Most have been developed for use in a wide range of industrial applications, although a few have been developed specifically for irrigation applications such as Rubicon SCADAConnect. SCADA is discussed in detail in Chapter 3.

The SCADA software has to be configured to meet site- and client-specific requirements. The different elements of the system that may need configuring include

- SCADA graphical user interface (GUI) overview screen and individual site screens, which display the important variables for each remote site (Fig. 8-5);
- Scaling of the data received from RTU/PLC if data need to be displayed in different units in the GUI;
- Services residing at the central server that are used to collect and store information arriving from the RTU/PLCs to the database;
- Polling of the sites as per client-specific requirements;
- Alarm thresholds for the different variables, alarm priority settings, alarm groups, recipient lists, alarm actions, and alarm escalation processes;
- Users, their roles, and authorizations;
- Reports; and
- Security.

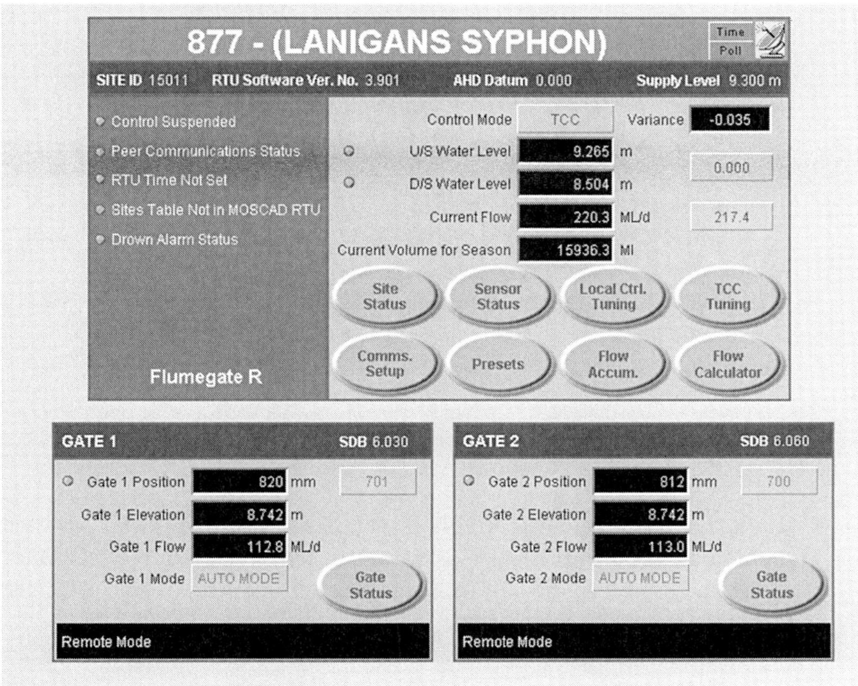

Fig. 8-5. Rubicon SCADAConnect GUI of a two-gate site displaying various process variables

It should be noted that the most basic SCADA systems only monitor field devices and cannot automatically control them. Automatic control functionality requires additional software in the RTU/PLC as well as water ordering and demand management software, or network visualization and control software, which are discussed in the following sections.

8.3.4. Water Ordering and Demand Management Software

Water ordering and demand management software provides a means for the irrigator to interact with irrigation authorities, view entitlements, view available capacity in the network, and place an order. The demand management component of the software ensures that a requested order does not violate the capacity and safe operation constraints of the canal network.

Water allocation and demand management software is intended to provide

- Water schedules at each irrigator outlet according to demand as approved by the irrigation authority;
- Flow schedules at each inline or head structure of the irrigation network;
- Ability to check that the schedule does not violate constraints such as
  - Water requirements of each order and their cumulative sum,
  - Maximum design discharge at each point in the network, and
  - Water level rate of change along the canal;
- A preview of the future demand to the underlying control system, which can be used in control computations; and
- Record keeping.

The demand can be estimated in several ways, including

- *Forecasting.* Forecasting involves predicting future water requirements based on inputs such as weather, crop type, previous statistical data, soil moisture measurements, or combinations of these factors.
- *Roster System.* Water is delivered to irrigators as per a scheduled roster that is set at the beginning of an irrigation season. The irrigation district personnel communicate with irrigators prior to their scheduled start to confirm the operation of their turnout or outlet.
- *Based on Irrigator Requests.* The irrigator places an order based on an estimate of crop requirements. This method has the largest potential for water savings but requires a flexible, on-demand service. Once an order is received, the scheduling can be carried out by a human or automatically by demand management software. The scheduling can be done in a variety of ways such as a simple first in, first out (FIFO) policy, or a more complex scheduling based on delivery shares and priorities. When placing an order, the irrigator must specify the
  - Outlet where the request is to be supplied,
  - Time and date of requested flow start or change,
  - Flow rate, and
  - Duration.

  Today's farmers have high expectations for flexibility and convenience, and all modern methods should be provided to place orders including
  - A Web browser interface that is available on the Internet or Smartphone;
  - Interactive voice response (IVR) technology (telephone and DTMF tones) using the public telephone system; and

  - A graphical user interface that has a direct connection to the central server via either the LAN or WAN and is updated by an operator of the irrigation authority.
- *Based on Actual Crop Needs.* Currently technology has advanced to the point that live measurements and knowledge of a crop type can be used to request a flow rate automatically for a time that exactly matches the crop requirements. The flow rate requested is a function of the farm topography, nature of soil, current moisture content of the soil, weather, and crop type. Unlike an irrigator's request for a *constant* flow rate based on the irrigator's estimate, this approach is scientific and can place a *time-varying* flow rate that is optimal for a crop's growth. Fundamental to this approach is the capability of the network control system to deliver the varying flow rate.

  For example, Rubicon's FarmConnect combined with Demand Management software (DMS) and network control system provides the capability to order and deliver water per the crop's actual needs to maximize crop yield, productivity, and profit, and at the same time minimize wastage of water and reduce pollution by preventing nutrient movement into aquifer or river systems.

To place orders, the operator requires appropriate authorization, which must be configured in the database. Additional checks are required to ensure that the time and date are legal, the order does not overlap with another order for the supply point, and the flow rates specified are within the specified static minimum and maximum flow rates for each individual supply point. In addition to these constraints, the system should be configured optionally to ensure the following:

- The customer has adequate volumetric entitlement to enable the request to be supplied.
- The total volume of water delivered does not exceed configured limits.
- The crop types to which the water is being supplied are recorded.
- Sufficient notice is provided. This notice period can be configured to as little as 1 h, depending on the physical characteristics of the system.
- The order start time does not exceed the preorder period.
- For the time period of the order, the capacity of all regulating structures between the supply point and the source of supply is not exceeded.
- The rate of change of flow with the network between the supply point and the source of supply is not exceeded. The rate of change can either be a heuristic dynamic upper-bound or bounds defined

by a hydraulic model that is programmed with the canal's operational constraints.

The accepted water order transactions reside in the database. In a completely automated setup, a preset service seamlessly translates the order transactions from the database and writes to the remote RTU that controls the site, when configured to do so. The RTU stores these prospective actions and executes them at the due time to make the delivery as requested. Alarms are raised if this transaction fails.

8.3.5. Network Representation Software

Network representation software provides the user with a topological representation of the irrigation network, with a user-friendly interface that allows irrigators to view the different components in the network, live information of the important variables such as water levels and flows, current demand in the network, and active alarms.

More importantly, the topological information provides the basis for the central server-based control software to perform network computations. SCADA software and its associated components such as alarm management and trending, and water ordering can be invoked through the network representation software. Network representation software is a "one-stop shop" for all the information and tools required by an irrigation authority to monitor and run an irrigation canal network.

An important customization job is to draw the schematics of the irrigation canal network, as seen in Fig. 8-6. The network schematic has to be configured to meet site- and client-specific requirements. The different elements of the system that may need customization or configuration include

- Icons to represent different objects on the network such as different types of control gates, farmer turnouts, and stock and domestic outlets;
- Attributes of each asset in the network such as maximum flow capacity, minimum flow, latitudes and longitudes, loss, and travel time;
- Variables to be displayed on the screen that represent real-time information;
- Pool boundaries for use by the control software to perform network computation;
- Area of the network under automatic control for the central server-based control software and demand management software;
- Capacity constraints in terms of rate of change of flow and recovery times for every pool in the network (This information is used by the demand management software to ensure that the orders are

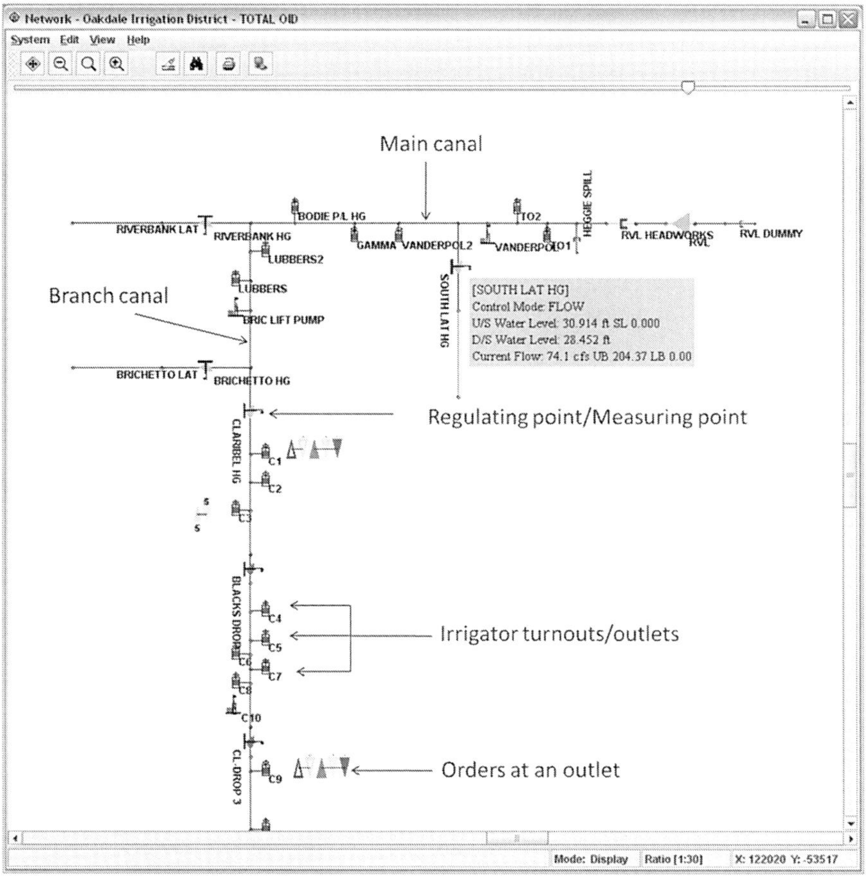

Fig. 8-6. Network visualization software
Source: Courtesy of Rubicon Water, Australia; reproduced with permission

scheduled such that the demand on the system never exceeds the physical limitations of the system); and

- Farmer turnouts that are to be operated by network control and demand management software. This is to ensure that a farmer order is scheduled automatically and flow is delivered automatically without any operator intervention, and irrigation is monitored constantly for noncompliance.

8.4. VERIFICATION OF INFRASTRUCTURE INSTALLATION AT FIELD SITES AND CENTRAL CONTROL ROOM

A number of other generic tasks under project delivery and management, such as the installation and verification of field devices, communications infrastructure, and IT infrastructure, are not included in this chapter because they are generally well understood. A good reference to project management is PMI (2008).

8.5. COMMISSIONING

After the control software, communications infrastructure, and control gates have been installed and configured, the components must be commissioned. *Commissioning* is the process by which equipment, a facility, or a plant is tested to verify if it functions according to its design objectives or specifications. At an irrigation control site, the control gates fitted with their instrumentation, communications hardware, and SCADA software must be tested to ensure that they can perform all the functions as required.

Many components of the site's infrastructure, such as the electronics, mechanical equipment, and some software features, are factory-tested and should be delivered after they already have been certified as ready to use. Therefore, this section focuses on the site commissioning *after* all of the different electronics and mechanical parts have been integrated and installed.

8.5.1. Commissioning of Control and SCADA Sites

This section only covers the commissioning of control gates that are actuated with electric motors, use ultrasonic sensors for water level measurement, utilize encoders for measuring gate position, and are controlled by an RTU/PLC. Similar guidelines can be developed for other types of gates that use different actuation and instrumentation.

Commissioning of the control gate is a standard procedure that may involve the following checks (in no specific order) after the site is powered up:

- Ensure that all connectors are pushed home in the controller housing properly and check for any loose cabling.
- Conduct a motor and encoder test. Move the gate in steps and ensure that the step size moved is exactly what was asked for.
- Ensure that the top left and right tip of the leaf of an overshot gate or radial gate are at the same level and likewise if the gate is an undershot, ensure that the bottom left and right tip of the gate are at the same level. A mismatch indicates that there is a tilt in the installation and will result in instrument damage or premature wear and tear. Follow similar strategies for other types of gates to ensure that the installation is perfect.
- Move the gate to its fully open and closed positions. The gate should be able to detect its fully open (upper) and fully closed

(lower) reference marks. If the reference mark is not detected, then a reference mark fail alarm should be triggered. This is to ensure that the motor doesn't drive the gate past its physical limits, which would damage the equipment.

- Check for encoder output when the gate is moving. If no encoder movement is detected during the testing, a "no gate movement" alarm should be triggered. This means that there are no signals being received from the encoder, and so no movement is detected while the motor is moving.
- Test the battery voltage, charging current, and site drain current. The variables should meet certain minimum criteria as required for the site.
- Verify that the battery temperature and controller housing temperatures that are obtained and displayed by the RTU/PLC are correct.
- Verify that the RTU/PLC telemetry module is communicating successfully with the gate's water level sensors, flow meters, or any other instrumentation attached, and confirm that the readings are correct.
- Test the communication with the network host. This communication may occur over a radio telemetry link, a wire line link, or through VSAT.
- Verify that there is enough signal strength at the site if radio telemetry is used.
- Calibrate the gate to a local site datum as per the irrigation authority standards.
- Check the various modes of operation of the site such as
  - *Local Manual*—the gate can be operated only locally through a human–machine interface (HMI) at the gate by an operator. The gate will not be available for control by an automatic controller on site and cannot be controlled remotely from the network host.
  - *Local Auto*—the gate is available for control by an automatic controller in the RTU/PLC but cannot be controlled or settings modified remotely from the network host.
  - *Remote Manual*—the gate is available for control remotely from the network host by an operator but is not available for control by an automatic controller on site or at the host.
  - *Remote Auto*—the gate is available for control remotely by an automatic controller on site or at the host, and the controller set points and modes can be changed through the host.

8.5.2. Commissioning of SCADA and Communications

Commissioning of SCADA and communications happens at the same time as the commissioning of the field equipment if the remote site is connected to SCADA. The following checks may be carried out during the commissioning of SCADA:

- Verify that the name and the communication ID correspond to the site that is being commissioned.
- Check that the SCADA screen contains all the gates and the variables to be displayed as required by the client.
- Verify that the alarm limits and alarm priorities are configured correctly for the different variables.
- Check that the supply level information and site data are set correctly.
- Verify that the readings in the remote RTU/PLC at the control gate match the readings on the SCADA GUI (graphical user interface).
- Check that the scaling of the readings obtained from the RTU/PLC to the corresponding display on the SCADA GUI is correct.
- Calibrate the deadband settings on flow and gate position per site- and pool-specific situations.
- Calibrate the "change of state" (COS) settings to meet the site-specific communication requirements.
- Verify that the user can communicate to the remote site through the SCADA GUI.
- Verify that the "event-based" communication occurs as per the COS settings.
- Verify that central server initiated polling occurs as scheduled.
- Verify that the user can move the gate remotely.
- Verify that the user can change the control mode at the site and that the site responds per the control objective.
- Verify that alarms are set by manually changing the trigger limits, and verify that alarm actions occur as configured.
- Verify that data received from the RTU/PLC are recorded in the database per the log settings of each variable.

Formal documentation of these commissioning procedures, also referred to as *inspection test plans* (ITPs) often form the basis for contractual acceptance.

8.6. CONTROL ROLLOUT AND TUNING

Once the commissioning of SCADA, remote sites, and communication is complete, the next step is to roll out or activate the control system. As discussed in Section 8.3, the objective may be to control a local variable at a site such as flow or the immediate upstream or downstream water levels, or there may be a global objective such as maintaining the water levels or volumes at set points throughout the network while reacting to the demand or supply. The global objective may also be linked to saving water by reducing spills and saving operations costs by reducing the need for buffer storages and reducing the electricity costs required for pumping.

Attainment of local objectives while satisfying the local requirements does not guarantee improved service for all irrigators at other parts of the network or operational efficiency, which are objectives that require a systemwide coordination. For the attainment of local objectives (at each gate) to achieve the global objective as well, it is essential that the former are aligned with the latter. To this end it is essential to deal with the complex issues of how all the control gates fit within the overall canal operation, how they are constrained by the canal hydraulics, and how they interact with the demand and satisfy the irrigators' requirements.

From this, it may be gleaned that it is best to carry out automation on a whole-of-canal basis rather than on a gate-by-gate basis. If carried out piecemeal, water savings at one location may be lost at another place, and the improvement in service and efficiency will at best be localized to the vicinity under automation. This may well defeat the global objective of better service to all the irrigators.

In this section, *rollout of control* refers to the control on a network scale as it encompasses the processes involved in rolling out control at just one location.

Rollout of control consists of the following steps:

1. Pool hydraulics are incorporated into an open-channel mathematical model, with its configuration based on the selected control strategy (e.g., immediate upstream or downstream level, distant downstream level, or volume). The model is calibrated for each pool in the canal network (see Chapter 5 for more details).
2. The controller is designed for feedback or feedforward control based on the mathematical model and selected control strategy (see Chapter 4 for details). Supervisory layers must also be designed, if necessary.
3. Components of the system are configured, including
   a. Feedback and feedforward controllers into the RTU/PLC (for local control) or at the central server (for centralized control);
   b. Supervisory control layer at the central server;
   c. Services residing in the central server, such as alarms, water ordering and demand management software, network control service, and others;
   d. Communications components of the software, such as peer communication addresses, change of state thresholds for selected process variables for each site, poll intervals, and others;

   e. Event logging in the database; and
   f. Alarm settings and alarm actions.
4. All components are implemented in the field.
5. Performance of the system is monitored to ensure proper operation.
6. All necessary components are fine-tuned.

8.6.1. Pool Hydraulics Characterization

Characterizing the pool hydraulics involves quantifying the dynamics of a system using mathematical models. Traditionally, open-channel hydraulics has been modeled by Saint Venant nonlinear hyperbolic partial differential equations (Chaudhary 1993). Saint Venant equations require accurate real-world data for calibration, and in practice these data are not always precisely known. Additionally, channel conditions can deteriorate after construction or after surveying. The Saint Venant equations also can be computationally very demanding for prediction and control purposes and must be simplified to be suitable for controller design. Another approach is to use models based on observed data from the system. Such an approach is referred to as *system identification* (Ljung 2007).

The model generated then enables the design of controllers using proven control theories to establish the stability and performance of the automatic control system. In the context of this chapter, the pool hydraulics characterization process involves the design and calibration of the chosen open canal model. Configuration of the model is a function of the selected control objective. The *configuration* of the models can start during early stages of the project once the control objective is known. However, *calibration* of the mathematical models requires data.

The data required to calibrate the Saint Venant equations can be obtained through

- Civil drawings of the channel construction ("as-built" drawings are best if available),
- Field surveys, and
- Comparison of simulation results with real-time data.

Data can be generated for the system identification approach to model calibration in either of two ways:

1. Generate the data through simulation of Saint Venant equations. Details of the approach are outlined in Ooi and Weyer (2008b). However, irrespective of how accurately and pedantically cross-section information has been gathered and used for the Saint Venant simulations, it is impossible to capture the entire dynamics of the

irrigation network. The mathematical models will be different from the real system by a certain time-varying, nonlinear degree.

2. Obtain data from the actual system under test conditions for a variety of operational regimes the canal system will experience (e.g., low and high flow regimes, and a variety of operating depths). This approach is commonly known as *step testing* in control engineering. The data from these tests can then be used to calibrate or validate the models. The data provide information such as travel time, time constant, disturbance tolerance capacity, and constraints on rate of change of flow. Organization of field tests requires coordination with irrigation authorities, synchronization with demand changes, availability of water, availability of instrumentation, and communications.

Conducting field tests for every pool in the canal network is not always possible, especially when controllers need to be designed for a large number of pools. Sometimes it can be challenging to align these tests with normal system operations of the irrigation canal. Experience shows that using data from offline hydraulic simulations using Saint Venant equations generally results in adequately calibrated models for controller design (Ooi and Weyer 2008b). Step tests only need to be conducted at selected locations to validate the models developed from offline physical simulations.

Based on the results of the conducted step tests at selected locations, control engineers will verify that the model developed from the physical data closely matches the actual system and make calculated adjustments as necessary. The engineers may scale the model for other sections of the canal where no geometry data exist or where it is not feasible to conduct step tests. Details of the system identification approach developed by Rubicon in conjunction with the University of Melbourne are published in Weyer (2001).

It is to be noted that in addition to the pool hydraulics, the gate discharge equations and gate servo mechanism must be calibrated to obtain accurate positioning or flow control.

8.6.2. Design and Configuration of the Controllers

The controller design process commences once the modeling exercise is complete. The process will involve deriving the controller parameters using *frequency response, state-space,* or *optimization* techniques based on the controller structure and design approach chosen to ensure that stability and performance criteria are met. Part of the controller design process that Rubicon developed in conjunction with the University of Melbourne is published in Weyer (2002) and Kearney et al. (2011a, 2011b).

Note that before field implementation, computer simulations of the network control will be carried out to fine-tune the controller parameters and assess the constraints of the automated irrigation network. Also a disturbance such as an irrigator starting or stopping at any point in the network will propagate through the network, and this has to be accounted for during the control design. The processes Rubicon uses have been developed in conjunction with the University of Melbourne and are incorporated in its controller design process, which is published in Ooi and Weyer (2008a), Li et al. (2005), and Kearney et al. (2011b).

8.6.3. Configuration

The configuration of controllers into the RTU or the central server will commence once the controller parameters are obtained after the steps mentioned in Section 8.5. The controller parameters can be entered into the RTU/PLC through the SCADA GUI, through the software toolboxes provided by the RTU/PLC supplier, or through the use of a script to bypass the SCADA GUI and write the parameters, if there is a large number of sites.

Also it is to be noted that in addition to the controllers, the host services, communication components, database, alarms, and alarm actions are configured and verified at this stage to enable automation operations.

8.6.4. Field Implementation

Activation of automatic control is best implemented on a top-down basis (i.e., activating automatic control at the control gate at the source of supply and progressively migrating one or more pools into network control in small manageable steps). This shifts the manual operation boundary downstream and constrains the negative effects of mistuned controllers. The pool controlled automatically is to be monitored for its performance and the controller fine-tuned if required. Artificial disturbance can be induced as well to test the performance of the control loops. Once deemed satisfactory by the irrigation district and the engineers, automation will be activated at more pools downstream until all the pools are under network control.

8.6.5. Control Performance Monitoring

As the pools begin operating under automatic control, control engineers will observe their behavior for some period to ensure that the designed controllers are behaving as expected. The engineers would like to see the canal operate over the full range of flow capacities and a variety of depths to observe controller response in all operating conditions. Sometimes the behavior of the actual system may differ from what has been modeled initially due to changing canal conditions such as aquatic growth or silt buildup. The controller may be too "aggressive," meaning that it is too fast for that particular pool or it may be too "sluggish," meaning that it is too slow to respond to system dynamics. Based on controller behavior, tuning may be required to adjust the controller to match the system dynamics better. Details of one of the approaches Rubicon developed in detecting unsatisfactory control behavior is described in Ooi and Weyer (2008a). Figures 8-7 and 8-8 show examples of this.

8.6.6. Fine-Tuning

As mentioned previously, sometimes the behavior of the actual system may differ from what has been modeled initially due to changing canal conditions such as aquatic growth or silt buildup, or sometimes the model may not have gone through the system's full range of operating conditions. Hence, fine-tuning of controllers may be required at intermittent stages. Controller tuning will be an ongoing process until the best set of controller parameters is determined. This may involve assessing the performance from the field data or carrying out more field tests.

During the process of control implementation and as the automated network grows, it may also be necessary to fine-tune the central server service's configuration parameters, enable more instances of the same host services to enable multitasking, fine-tune communication parameters, fine-tune alarm settings, configure, and other activities. This is an ongoing process that lasts from the beginning of automation activation all the way to the performance assessment and acceptance phase (see Section 8.5) and through to the operations and maintenance phase.

A summary of the pool hydraulics characterization, controller design, and tuning process is shown in Fig. 8-9.

8.7. PERFORMANCE ASSESSMENT AND ACCEPTANCE

As important as setting the overall project objectives are the measurement of the performance of the canal system against those objectives.

Once every element of the control solution is implemented in the field, controllers are tuned and the system is operational, a system performance acceptance test is conducted to ensure that the client's operational and management expectations have been met and that the vendor that delivered the project is paid.

The system performance acceptance test is usually a customized test that varies from one irrigation district to another. Irrigation authority staff

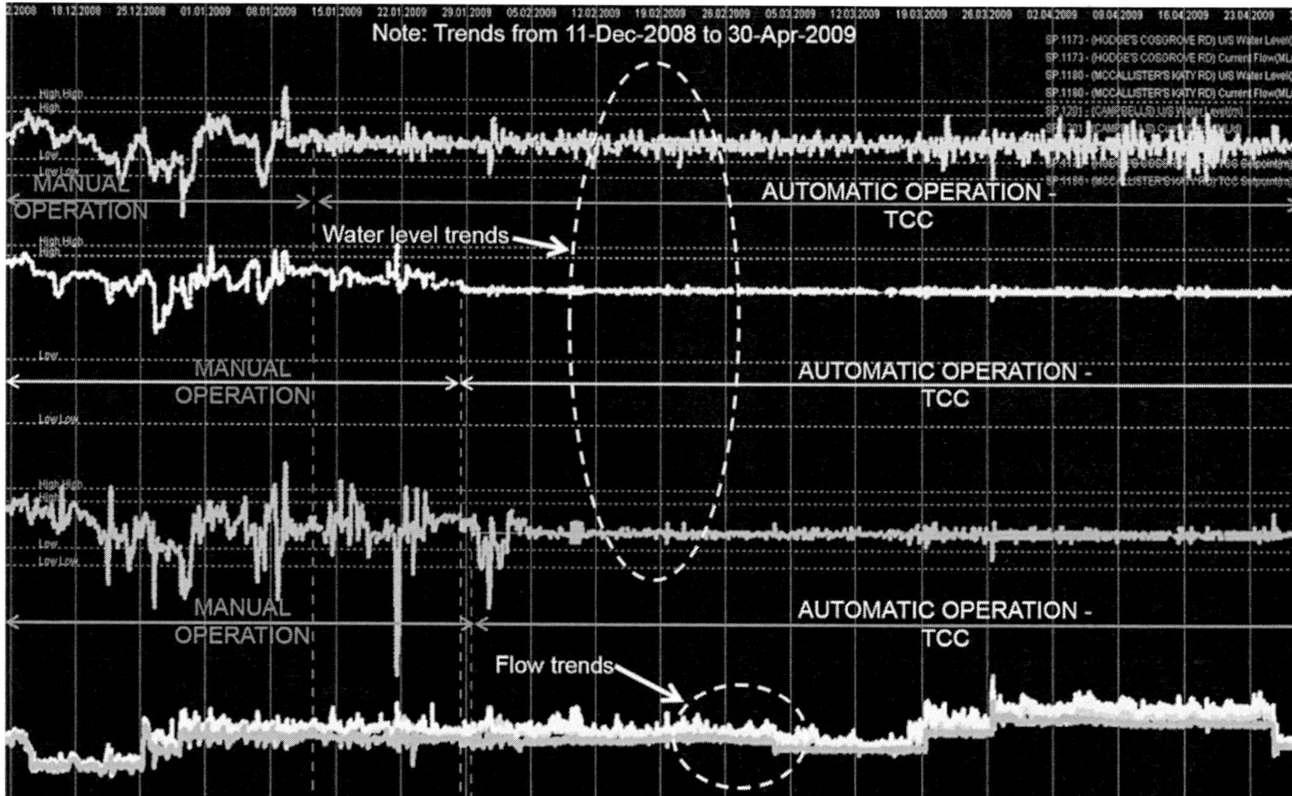

Fig. 8-7. Trends of water levels and flow prior to and after Rubicon's network control technology, also known as total channel control (TCC), was activated at East Goulburn Main Channel, G-MW, Shepparton, Australia
Source: Courtesy of Rubicon Water, Australia; reproduced with permission

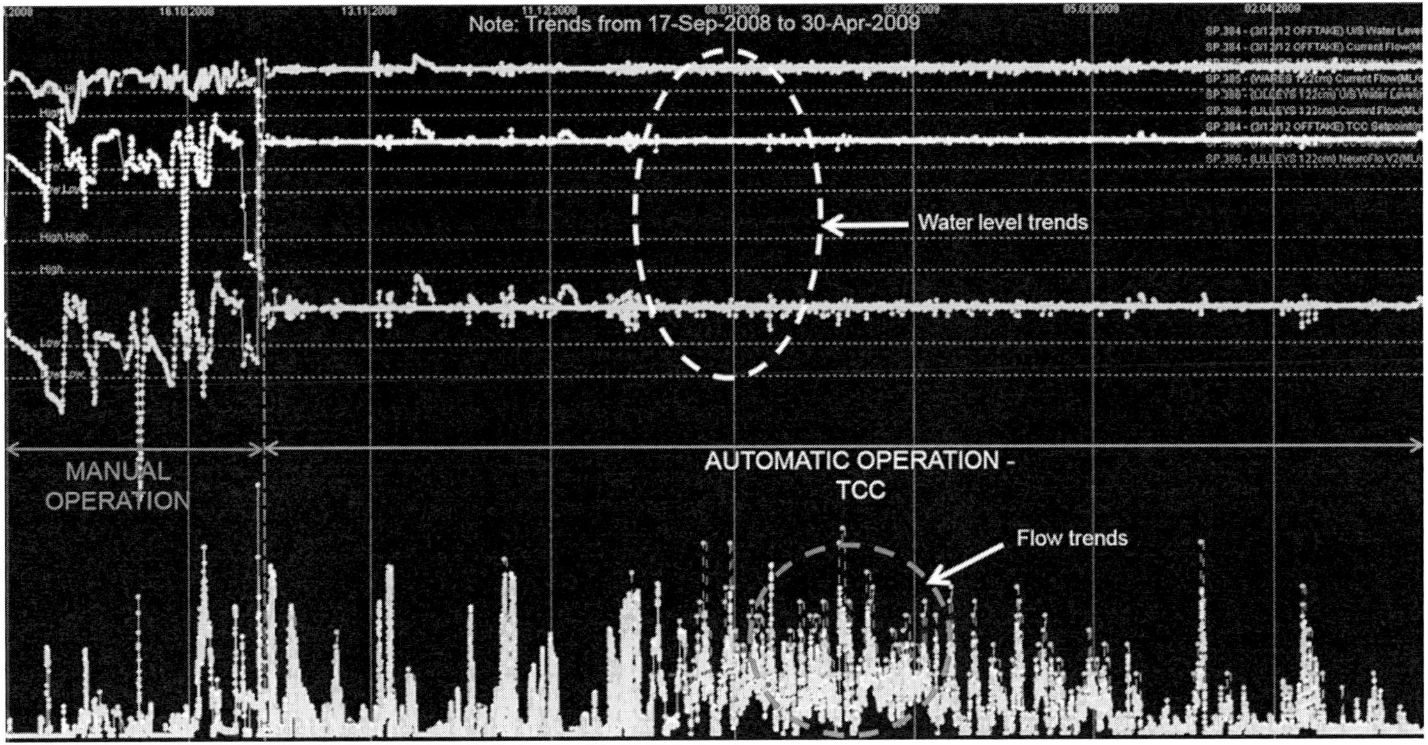

Fig. 8-8. Trends of water levels and flows prior to and after Rubicon's network control technology, also known as total channel control (TCC), was activated at EG 3/12/12 channel, G-MW, Shepparton, Australia
Source: Courtesy of Rubicon Water, Australia; reproduced with permission

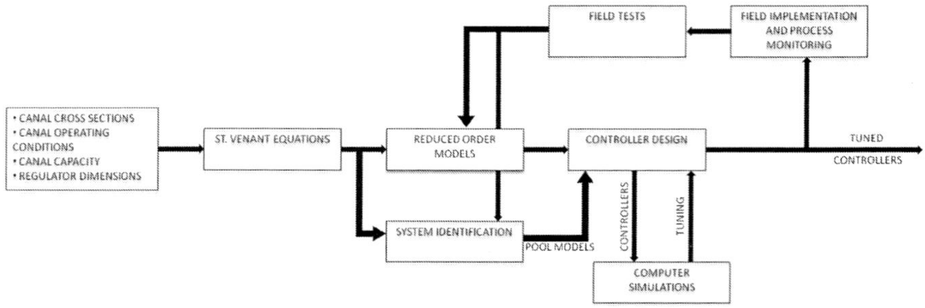

Fig. 8-9. Pool hydraulic characterization, controller design, and controller tuning process

generally find the methods to assess the ability of a system to reject disturbances documented in the preceding chapter impractical. For this reason, three simple tests have found favor in Australia and are discussed as follows:

1. Water level alarm performance metric,
2. Outfall volume performance metric or administrative spill volume performance metric, and
3. Successful irrigation index (SII).

8.7.1. Water Level Alarm Performance Metric

In this metric, a canal was deemed to be performing satisfactorily if the cumulative time that the upstream water level alarm is active was no more than 7.5% for a chosen period. This is a good metric to use where project objectives are concerned particularly with maintaining stable water levels. This computation was undertaken per channel or part thereof and can be computed as Eq. (8-1):

$$\frac{\sum \text{Time in alarm for each site}}{\text{Total time}} \tag{8-1}$$

The alarm limits are defined during the software configuration or the SCADA commissioning process and are uniquely configured for each pool. An example of the water level alarm performance report is shown in Fig. 8-10.

8.7.2. Outfall Volume Performance Metric

In the *outfall volume performance* metric, the cumulative outfall volume (or spillage at the tail of the canal) during the two-week acceptance period was to be no more than 0.05% of the inflow into the canal system (except

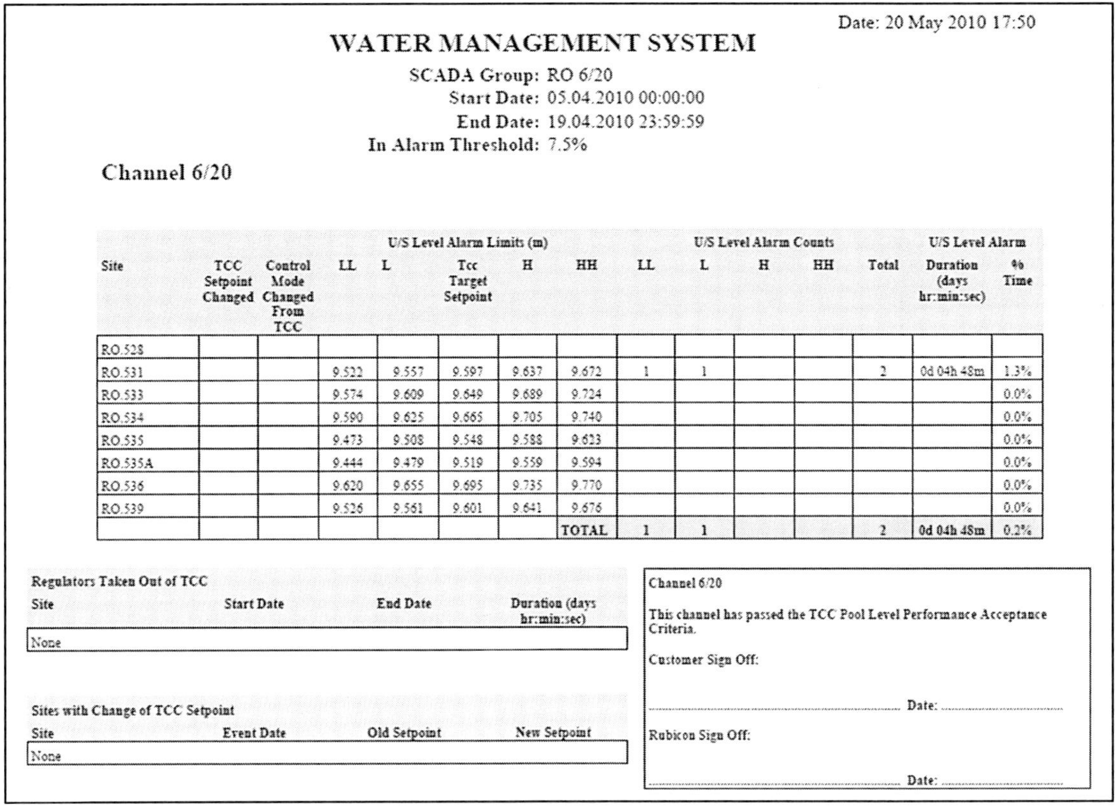

Date: 20 May 2010 17:50

WATER MANAGEMENT SYSTEM

SCADA Group: RO 6/20
Start Date: 05.04.2010 00:00:00
End Date: 19.04.2010 23:59:59
In Alarm Threshold: 7.5%

Channel 6/20

Site	TCC Setpoint Changed	Control Mode Changed From TCC	U/S Level Alarm Limits (m) LL	L	Tcc Target Setpoint	H	HH	U/S Level Alarm Counts LL	L	H	HH	Total	U/S Level Alarm Duration (days hr:min:sec)	% Time
RO.528														
RO.531			9.522	9.557	9.597	9.637	9.672	1	1			2	0d 04h 48m	1.3%
RO.533			9.574	9.609	9.649	9.689	9.724							0.0%
RO.534			9.590	9.625	9.665	9.705	9.740							0.0%
RO.535			9.473	9.508	9.548	9.588	9.623							0.0%
RO.535A			9.444	9.479	9.519	9.559	9.594							0.0%
RO.536			9.620	9.655	9.695	9.735	9.770							0.0%
RO.539			9.526	9.561	9.601	9.641	9.676							0.0%
							TOTAL	1	1			2	0d 04h 48m	0.2%

Regulators Taken Out of TCC

Site	Start Date	End Date	Duration (days hr:min:sec)
None			

Sites with Change of TCC Setpoint

Site	Event Date	Old Setpoint	New Setpoint
None			

Channel 6/20

This channel has passed the TCC Pool Level Performance Acceptance Criteria.

Customer Sign Off:

.... Date: ....

Rubicon Sign Off:

.... Date: ....

Fig. 8-10. Screenshot of water level alarm performance acceptance report for RO6/20 canal, Goulburn-Murray Water, Victoria, Australia
Source: Courtesy of Rubicon Water, Australia; reproduced with permission

during a major shutdown where the ordered demand reduces to zero). This is an effective metric when water savings are an important project objective. An example of outfall volume performance metric is shown in Fig. 8-11.

8.7.3. Successful Irrigation Index

A metric called the *successful irrigation index* (SII), now called *successful delivery index* (SDI), developed by Southern Rural Water in conjunction with Rubicon in Australia (SRW 2008), provides a measure of delivery reliability. SII categorizes an irrigation delivery as a success or failure on the basis of whether the delivered flow was within ±5% of the ordered flow or ±0.5 ML/day (whichever is greater) for 95% of the order duration. Figure 8-12 shows a SCADA trend of flows delivered to farmers for every month of the 2007–2008 irrigation season at Southern Rural Water and the respective SII values. Out of a total of 764 irrigation deliveries, 84% were classified as "successful."

Armed with this definition, root cause analysis can be undertaken of the "unsuccessful" irrigations to provide pointers to improving system-wide performance. In the case cited here, it was found that all of the unsuccessful irrigations were caused by problems at the demand scheduling level, poor customer infrastructure (generally high ground on farm or undersized on-farm infrastructure), and incorrect manual intervention during automated operation (Figs. 8-13 and 8-14).

8.8. POST-ACCEPTANCE OPERATION AND MAINTENANCE (O&M)

Post-acceptance operations and maintenance is an important part of canal automation. It is the analog of a routine maintenance program that irrigation districts generally undertake anyway but includes new equipment like the software and operation.

The irrigation district can expect manufacturers' warranties on the field instrumentation, communications infrastructure, and computer hardware. Depending on the contract with the automation contractor, the irrigation district can negotiate a comprehensive warranty for a certain period. The contractor particularly may be asked to provide warranty on all supplies and services for this period, including all required spares, replacements, and hardware, or to even provide the replacement services. It can also include

- The presence of expert on-site staff who are responsible for O&M of hardware, software installed, and electromechanical components and

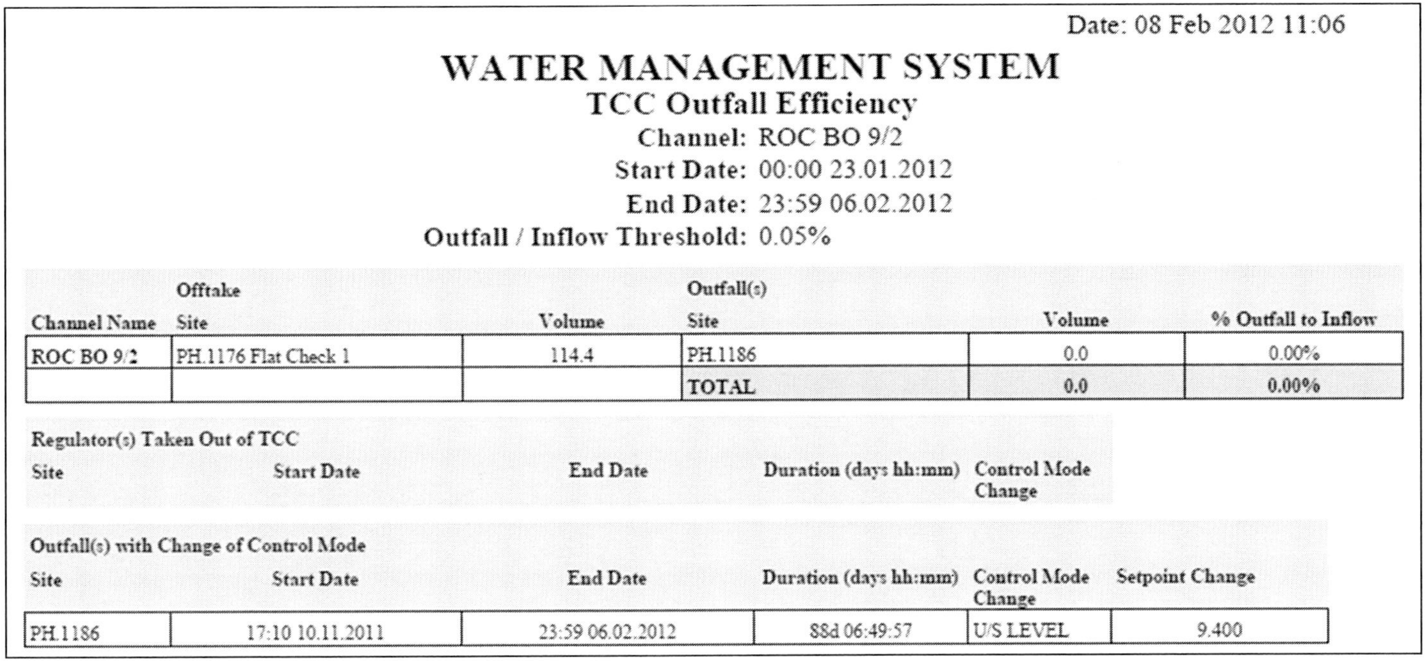

Date: 08 Feb 2012 11:06

WATER MANAGEMENT SYSTEM

TCC Outfall Efficiency

Channel: ROC BO 9/2
Start Date: 00:00 23.01.2012
End Date: 23:59 06.02.2012
Outfall / Inflow Threshold: 0.05%

	Offtake		Outfall(s)		
Channel Name	Site	Volume	Site	Volume	% Outfall to Inflow
ROC BO 9/2	PH.1176 Flat Check 1	114.4	PH.1186	0.0	0.00%
			TOTAL	0.0	0.00%

Regulator(s) Taken Out of TCC

Site	Start Date	End Date	Duration (days hh:mm)	Control Mode Change

Outfall(s) with Change of Control Mode

Site	Start Date	End Date	Duration (days hh:mm)	Control Mode Change	Setpoint Change
PH.1186	17:10 10.11.2011	23:59 06.02.2012	88d 06:49:57	U/S LEVEL	9.400

Fig. 8-11. Screenshot of outfall efficiency report for BO 9/2 canal, Goulburn-Murray Water, Victoria, Australia
Source: Courtesy of Rubicon Water, Australia; reproduced with permission

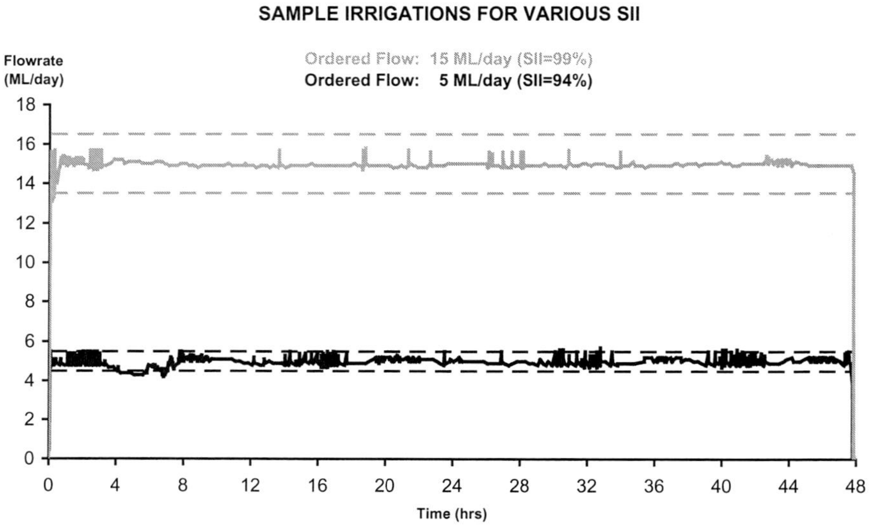

Fig. 8-12. Plot showing the trends of the flow delivered to a farmer and the SII
Source: Courtesy of Rubicon Water, Australia; reproduced with permission

- A detailed O&M plan describing the O&M activities and taking into consideration manufacturer or OEM recommendations for all hardware and software.

With respect to software and operations, a key tool is what is commonly called a "support and maintenance agreement" with the software suppliers. Such agreements can provide

- Operations support (e.g., a telephone help line);
- Preventive maintenance of the automation system, including software bug fixes or upgrades; and
- Corrective maintenance of the automation system, including tuning services or more training for new staff.

Where an irrigation district has contracted for the supply of the entire automation project, it has the additional possibility to leverage the skills and resources of the suppliers to achieve systemwide performance goals. If the irrigation district can express its operations in terms of high-level outcomes, such as those discussed in Section 8.6, the supplier can be contracted to maintain such outcomes. This is common in many other areas of engineering and provides a new and powerful tool for irrigation districts. It is important to note that the cost of post-implementation operations and maintenance has to be considered in the initial project

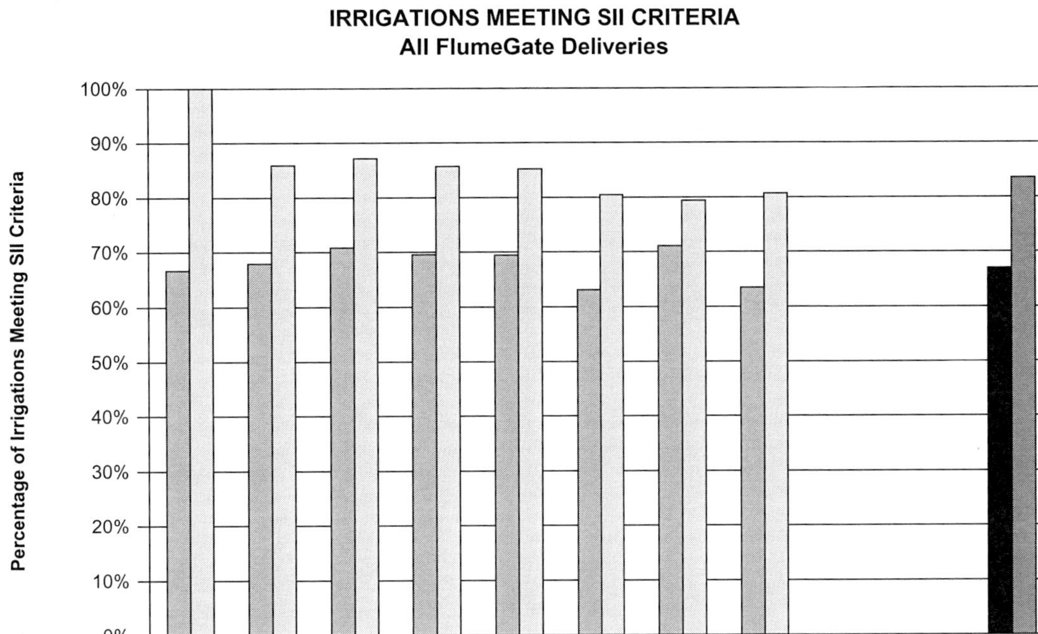

	Aug-07	Sep-07	Oct-07	Nov-07	Dec-07	Jan-08	Feb-08	Mar-08	Apr-08	May-08	Season to Date
% of irrigations which met SII criteria	67%	68%	71%	70%	69%	63%	71%	63%	0%	0%	67%
% of irrigations which met SII criteria excluding irrigations affected by customer side issues	100%	86%	87%	86%	85%	80%	79%	81%	0%	0%	84%
Total number of irrigations	3	81	144	69	108	176	38	145	-	-	764

Fig. 8-13. Monthly breakdown of successful irrigation index of all deliveries by total channel control (TCC) at Southern Rural Water

Source: Courtesy of Rubicon Water, Australia; reproduced with permission

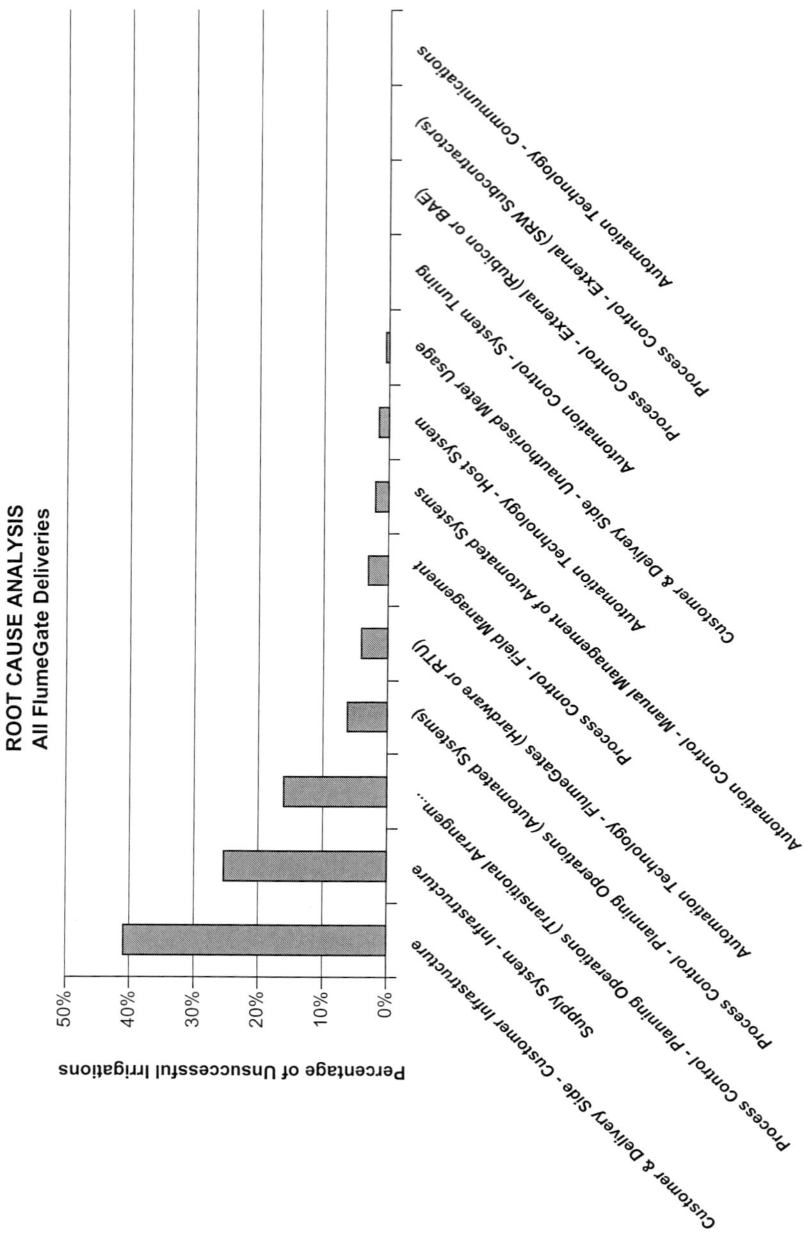

Fig. 8-14. Root cause analysis of unsuccessful irrigation at Southern Rural Water
Source: Courtesy of Rubicon Water, Australia; reproduced with permission

budget. Details of various successful canal automation projects can be found in Rubicon Water (2013).

8.9. TRAINING

An automated canal system requires that operators learn a number of new computer skills and adapt to a style of operation that involves reacting to operational problems rather than simply implementing routines (Fig. 8-15). This is similar to the role of a pilot when an airplane is operating in autopilot. The pilot must monitor instruments vigilantly and correct for unforeseen circumstances or deal with alarms.

Initially, operators may be fearful of change and unsure of what it means for their jobs. Yet, in most cases, the changes almost always result in positive outcomes. Not only do operators improve their skills and

Fig. 8-15. Operators learn to set flow set points at a gate rather than to adjust the height of a gate leaf
Source: Photo courtesy of Michael Kai, www.michaelkai.net; reproduced with permission

knowledge, but their jobs become more varied and productive. Nonetheless, irrigation district managers need to manage the transition with care, and training is an integral part of that process.

A common challenge for operators is learning to read and interpret data and graphs on a computer screen rather than the visual, visceral experience of walking along a canal bank. A second, perhaps more important challenge is the change from an intuitive mode of thinking and problem solving based on experience and rule-of-thumb intervention to an acceptance of the realities of what their sensors are telling them. Prior to the advent of RTU/PLCs, SCADA, and telemetry, the canal operators lacked information about the effect of their interventions on other parts of the channel network. Early in the operational life of an automated system, when operators try to apply those rules of thumb that have served them so well, they are often surprised at how their interventions may disrupt the automated operation in other parts of the network (see Fig. 8-3).

Canal automation does not just affect operation. New equipment, such as radios or automated gates, requires different maintenance routines and maintenance skills. Again, the new skills provide greater job satisfaction and can make the irrigation district an attractive employer for younger employees.

Training should be conducted at different phases of the project as shown in Fig. 8-2 to ensure that the irrigation district personnel are not fearful of the change, develop the new skills to operate the automated system, take preventive actions, and troubleshoot problems.

Training should cover the following:

- Control gate and sensor operations and maintenance;
- Communication system operations and maintenance;
- SCADA operations; and
- Control software operation, including network control, water ordering, and demand management.

A virtual mock environment of the real SCADA system can be set up easily for training purposes.

8.10. MANUALS

In addition to the training, the irrigation district must be supplied with manuals for every component that makes up its system. Manuals serve as a reference to the irrigation district personnel and the engineers in case they need to troubleshoot, enhance existing functionalities, completely revamp the existing system, extend the scope of the functionalities of the existing hardware and software, or integrate the existing system with a third-party system.

The automation contractor should supply technical manuals for each item of the SCADA software, RTU/PLC software, water ordering and demand management software, network representation software, RTU/PLC hardware, communications hardware and architecture, control gates, and instrumentation.

8.10.1. System Description Manuals

System manuals should contain an integrated description of the SCADA hardware and software, including, at a minimum, the following information:

- *System overview* to provide architecture of the system implementation, as well as a general understanding of the hardware, software, control functions, interfaces, and general data flow associated with the entire system; the system overview document should include a block diagram of the entire system;
- *Hardware system description* should detail each major hardware subsystem; at a minimum, descriptions shall include central server infrastructure, RTU/PLC infrastructure, and communications network infrastructure; and
- *Software system description* describing the general software design of the total system; functional block diagrams should be included with detailed descriptions on a subsystem-by-subsystem basis, and a top-level diagram should show the functional relationships of the major control software components, and the water allocation and demand management components.

8.10.2. Hardware Manuals

The hardware documentation should be in sufficient detail to provide information to install, operate, expand, and add new equipment, as well as to provide information for preventive and corrective maintenance of the automated system. The hardware documentation should at a minimum include the following information:

- Theory of operation;
- Installation instructions;
- Maintenance procedures, including troubleshooting only;
- Test equipment;
- Schematic and logic diagrams; and
- Wiring and cable diagrams.

8.10.3. Software Manuals

The software documentation should be in sufficient detail to provide information to install, operate, expand, change, troubleshoot, and add new software modules and programs, as well as to provide information for preventive and corrective maintenance of the SCADA system software. The software documentation should at a minimum include the following information:

- A *software subsystem design document* that extends the summary descriptions for each software subsystem found in the system description document and includes relationships between each subsystem and other software subsystems. All functions performed by the software should be described and linked to the programs that perform these functions. The software subsystem design document should describe the following minimum information:
  - RTU/PLC software
  - Communications software
  - SCADA software
  - Water ordering and demand management software
  - Network visualization software.
- *Software detailed design documents* for every software system. The descriptions should be in sufficient detail to enable coding and testing of each program. This includes:
  - Detailed study of the functions and the components developed to implement them
  - Detailed descriptions of each database structure, and the format for all data used by the database; a description shall be provided for each database in the system, and the document should include descriptions for all data structures referenced by programs inputting data to or outputting data from the real-time system databases and others
  - Description of the interfaces provided by the software.

8.10.4. Operator User Manual

This document should contain a comprehensive description of the operation of all furnished software. This document should be provided as a guide to the operator and should contain the operating instructions for the control gates, SCADA, network visualization, and water ordering and demand management software. It should include all procedures necessary for an operator to operate the system independently. Illustrations of the equipment should be used wherever possible to aid operators.

8.11. REFERENCES

Chaudhary, M. H. (1993). *Open channel flow*. Prentice Hall: Englewood Cliffs, NJ.

Kearney, M., Cantoni, M., and Dower, P. M. (2011a). "Model predictive control for systems with scheduled load and its application to automated irrigation channels." *Proc., IEEE Int. Conf. Networking, Sensing, and Control*, Delft, The Netherlands, 186–191.

Kearney. M., Cantoni, M., and Dower, P. M. (2011b). "Non-iterative distributed MPC for large-scale irrigation channels." *Proc., Australian Control Conf.*, Melbourne, Australia, 217–223.

Li, Y., Cantoni, M., and Weyer, E. (2005). "On water-level error propagation in controlled irrigation channels." *Proc., 44th IEEE Conf. Decision and Control, and the European Control Conf., CDC-ECC*, Melbourne, Australia, 2101–2106.

Ljung, L. (2007). *System identification: Theory for the user*. Prentice Hall: New York.

Motorola. (2007). "SCADA systems white paper." <http://www.motorolasolutions.com/web/Business/Products/SCADA%20Products/_Documents/Static%20Files/SCADA_Sys_Wht_Ppr-2a_New.pdf> (June 19, 2013).

Ooi, S., and Weyer, E. (2008a). "Detection of oscillatory control loops in irrigation channels." *Proc., 17th IFAC World Congress*, COEX, Seoul, South Korea, 5500–5505.

Ooi, S., and Weyer, E. (2008b). "Control design for an irrigation channel from physical data." *Control Eng. Pract.*, 16(9), 1132–1150.

Project Management Institute. (PMI). (2008). *A guide to the project management body of knowledge (PMBOK guide)*, 4th Ed., Newtown Square, PA.

Rubicon Water. (2013). "Customer stories." <http://www.rubiconwater.com/catalogue/usa-customer-stories> (June 19, 2013).

Southern Rural Water. (SRW). (2008). *Macalister irrigation district channel automation project—Performance reporting*. Maffra, VIC, Australia.

Weyer, E. (2001). "System identification of an open water channel." *Control Eng. Pract.*, 9, 1289–1299. Based on a prior paper in *IFAC Symposium on System Identification*, 2000.

Weyer, E. (2002). "Decentralised PI control of an open water channel." *Proc., 15th IFAC World Congress*, Barcelona, Spain.

APPENDIX A

GLOSSARY OF CANAL AUTOMATION TERMINOLOGY

Accumulator—A method to keep track of or store requested control actions (or hydraulic pressure) that cannot be implemented due to physical limitations (e.g., minimum gate movement), often used to preserve the integral action of a controller.

Actuator—Device used to physically move a piece of hardware, such as a gate. In general terms, it converts the control output to a mechanical operation that affects the process being controlled.

Adaptive Control—A control method whereby various parameters that define the controller are adjusted automatically by the controller itself based on prior control performance; can be applied to a variety of control methods.

Adjoint—A solution method in operations research where, mathematically, the meaning of the physical variables and the constraints are reversed (typically one adjoint variable per constraint). For linear systems theory, it can be shown mathematically that the solution of the adjoint problem can be used to solve the original problem with less computational difficulty.

Algorithm—A prescribed set of well-defined rules or processes for the solution of a problem in a finite number of steps, usually expressed in the form of Boolean logic and mathematical equations.

AMIL—Hydromechanical automatic gates developed by Alstom (formerly GEC Alsthom) for controlling the water level immediately upstream of the gate.

Analog—A continuously variable electrical signal representing a measured quantity (e.g., electrical signals such as current, voltage, frequency, or phase used to represent physical quantities such as water level, flow, and gate position).

Analog-to-Digital (A-to-D) Converter—An electronic device that converts an analog signal to a digital equivalent (e.g., binary or hexadecimal).

Antenna—A device that sends and/or receives radio signals. Antennas can be unidirectional (focused in a specific direction, e.g Yagi) or omnidirectional (not focused).

Anti-Hunt Device—A system used to dampen the response of a controller to prevent oscillations that would cause excessive control actions.

Anticipatory Open-Loop Control—An open-loop control based on prior knowledge of needed flow changes, future disturbances, or changes in set points (tracking references).

Arranged Delivery—See **Scheduled Delivery**.

Auto-Regressive Integrated Moving Average State Model (ARIMAX)—A model that fits a time series of data to this type of statistical time-series model for describing the transition from one state to another.

Automatic Control—A procedure or method used to regulate mechanical or electrical equipment without human observation, effort, or decision.

Automation—A procedure or method used to regulate a water system by mechanical or electronic equipment that takes the place of human observation, effort, and decision; the condition of being automatically controlled.

AVIS-AVIO—Hydromechanical automatic gates developed by Alstom (formerly GEC Alsthom) for controlling the water level immediately downstream of the gate (AVIS for small and medium head losses; AVIO for high head losses).

Backwater Surface Areas—The water surface area of the canal pool upstream from a control structure that is influenced by the effects of backwater from the structure.

Bandwidth of the Frequency Response—The range of resonance frequencies over which control is effective. For larger bandwidth, the response is faster; however, the response is more sensitive to noise and parameter variation and is less robust.

Bifurcation—A junction in which the supply flow splits into two downstream paths.

Binary—A number expressed in base 2, which is represented as a series of ones and zeros.

Binary Coded Decimal (BCD)—Binary representation using four binary bits to represent decimal numbers 0 through 9. Remaining binary combinations are not used, and a new set of four is used for the next decimal number.

Black Box Model—A model of a system where the inputs are related to the outputs with a model that does not make intuitive sense or is not based on a technical understanding of the physical processes involved.

Block Diagram—A diagram of a system used to define relationships between disturbances (w), control actions (u), controlled variables (z), and measurements (y). These diagrams are used to define and solve control problems.

Bode Plots or Response—In control theory, a plot of the magnitude (and/or phase) of the output response versus frequency of input signal. These plots determine both the resonance frequency and the magnitude of the associated output.

Boundary Conditions—Flow conditions imposed at the ends of a pipeline or canal reach by various physical structures, which must be described mathematically to solve the general equation of flow for unsteady-flow computer models.

Bounded Variation of Parameters—The range of values for a parameter or physical variable.

Broad-Crested Weir—A measuring weir where the length of the crest, in the direction of flow, is roughly twice as long as the upstream head. (Streamlined broad-crested weirs have similar hydraulic properties as long-throated flumes.)

Buffer Reservoir—A reservoir used in canal systems to reduce the mismatch between downstream demands and upstream water supplies to maintain a balanced operation; also known as a **Regulation Reservoir**.

Canal Automation—The implementation of a control system that upgrades the conventional method of canal system operation.

Canal Automation for Rapid Demand Deliveries (CARDD)—A heuristic gate control algorithm developed by Charles Burt, Cal Poly, San Luis Obispo, CA.

Canal Check Gate Structure—A structure designed to control the water surface level and flow in a canal, maintaining a specified water depth or head on outlets or turnout structures. Most canal check structures have movable gates.

Canal Frequency Response—The output response of a canal as a function of the frequency of inputs or disturbances, which can be shown on a Bode plot, a key feature of which is the resonance frequency.

Canal Pool—Canal section between check structures.

Canal Prism—The cross-sectional shape of a canal.

Canal Reach—Segment of main canal system with a uniform cross section. Canal reaches can include drops or short pipe sections, as long as the canal cross section is the same upstream and downstream. A canal reach can include several canal pools or can be one segment within a canal pool.

Cascade Control—A method of automation using control units arranged in sequence such that a given unit is controlled by the preceding unit and controls the following unit.

Celerity—The speed of propagation of a gravity wave, for example, the speed of ripples on a pond.

Central Control Center—A location where control actions can be made at many other remote sites, typically at a strategic location.

Centralized Control—Control of a canal project from a central location generally by a master station, communications network, and one (or more) **Remote Terminal Unit** (RTU).

Centralized Proportional-Integral (PI) Controller—A controller where all gates are adjusted based on all measured water levels. As defined, it also includes the influence of prior control actions (i.e., a **Smith Predictor**).

Check Gate—A gate located at a check structure used to control flow.

Classical Control Techniques—A series of control methods based on linear systems theory, typically the **Derivative Feedback Control, Integral Feedback Control**, or **Proportional Feedback Control**, or combinations thereof.

Classical Nyquist Control Concepts—The performance of classical control methods can be described by Nyquist plots. These plots can determine whether controllers are stable or unstable and indicate robustness margins (range of conditions over which control is robust).

Classical Regulators—A controller or regulator based on classical control theory.

Closed-Loop System—A system where the input (control action) is adjusted automatically (manipulated) based on the measured output.

Communication Channel Type—Simplex (one direction), half duplex (both directions but only one at a time (switchable), and full duplex (both directions concurrently).

Communications Protocol—A method for sending and receiving data between two electronic devices.

Computation Time Step—The time step over which water flow is determined in unsteady-flow models.

Conditional Gradient Method—A gradient-search method whereby the direction and magnitude of the next guess in a search for an optimal solution is limited by user-defined conditions (e.g., by the range of possible values or the amount of change allowed between guesses).

Constant-Discharge Regulator—A control device or method that provides local, constant flow.

Constant Volume Operation Method—A canal operation that maintains a relatively constant water volume in each canal pool.

Contour Evaluation—See **Polar Plot**.

Control Action—The change in a device or condition that influences the output of the system of interest.

Control Cost Weighting Matrix—A matrix that describes the cost or penalty associated with a controlled variable (e.g., water level) that deviates from the set point (e.g., water level error) or with a control action (e.g., gate movement).

Control Element—A part of a control system through which the system's process is regulated.

Control Horizon—Time span over which deviations from the set point are considered by the controller; commonly used in predictive control.

Control Interval—The interval of time between successive control actions.

Control Model Robustness—A measure of the ability of a control method to remain stable and effective over a wide range of conditions.

Control Scheme—The collection of methods and algorithms brought together to accomplish control of a canal system.

Control Sequence—Actions for control that are taken one after the other.

Control System—An arrangement of electronic, electrical, and mechanical components that commands or directs the regulation of a canal system.

Control (Action) Variable—Describes what is changed when a control action takes place (i.e., the input variable).

Controllability—A measure of whether stable control can be obtained for a particular physical system.

Controlled Variable—What is to be controlled to a particular value (i.e., the output variable).

Controlled Volume Method, Concept, or Operation—A method for maintaining or controlling the volume in a canal pool or a series of canal pools.

Controller Coefficients—Values used to determine a new numerical value of the control action from the current numerical values of system outputs (measurements) and/or from prior conditions.

Controller Gain Matrix—A matrix that contains controller coefficients.

Controller State Variables—The variables that define the current state of the system for control purposes.

Corrective Action—An action taken to bring a controlled system back to the desired state.

Cost Function—See **Control Cost Weighting Matrix**.

Coupling Effects—The result of the physical connection between adjacent systems; in canal control, the hydraulic influence of one canal pool on adjacent canal pools.

Damped System—A closed-loop system whose response aggressiveness has been lessened. Damped systems tend to be slower reacting but more stable.

Damping—The degree to which a system is damped.

De Saint-Venant Equations—See **Saint Venant Equations**.

Deadband—The range through which the measured signal can vary without initiating a control action.

Decoupler I—A method for removing the hydraulic influence of one canal pool on all upstream canal pools.

Decoupler II—A method for removing the hydraulic influence of one canal pool on the canal pool immediately downstream and, thus, on all downstream pools.

Decoupling—A method for removing the hydraulic interactions between adjacent canal pools.

Decrement—The change in the output (e.g., water level) for a change in the state of the system (e.g., canal flow rate).

Delay Time—The time between a control action upstream and the resulting change downstream.

Delivery Flexibility—The flexibility that water users have in requesting delivery changes and the ability of the canal system to accommodate the request.

Delivery System—A series of canals and pipelines that convey water from a single source, such as a storage reservoir, to a number of individual points of use. The delivery system is a common classification that is associated with canal systems for irrigation, municipal and industrial, and fish and wildlife needs.

Demand-Oriented System—An irrigation water distribution system (canal or pipeline) that attempts to accommodate user demands, often where control actions are based on conditions/demands downstream.

Dependence Domain—The range of conditions over which system components influence each other.

Derivative Feedback Control—A method whereby control actions are based on the rate at which the output is moving relative to the set point.

Digital—Representation of a quantity by an arrangement of digits, each of which represents a portion or weighted portion of the quantity, for example, in communications, transmission of quantities by a series of pulses.

Digital Control—A system whereby control actions are taken at a discrete time interval based on digital measurement of system conditions.

Digital Measurement—A measured quantity that is expressed as a digital number, often as the result of conversion from analog signals by an **Analog to Digital (A-to-D)** conversion.

Digital, State-Feedback Control—A system based on measurements and control actions at discrete times where the physical system and how it changes from one discrete time to the next is described by a series of variables called "states" and by a matrix that defines the transition from one state to another.

Discrete Difference Form—For controllers, a form in which all variables are described by their change in value from one discrete time to another; also called **Incremental Form** or **Velocity Form**.

Discrete-Time Linear Quadratic Regulator (LQR)—A control method that is based on a physical system, which can be described in state-space form, and whose control parameters are determined by the minimization of quadratic performance criteria on controlled variables and control action variables.

Discretization—The process of converting a system of continuous variables into a system of discrete variables.

Distribution System—Delivers water from the main canal-side turnout to individual water users or to other smaller distribution systems.

Disturbance-Wave Travel Time—The time for a change in flow rate (disturbance) to travel from one point to another (e.g., from one end of a pool or canal to the other).

Disturbances—Changes in inflow and outflow to a canal or canal pool.

Ditchrider—Canal system operations personnel; the person responsible for controlling the canal system on site based on the flow schedule established by the water master.

Diversion Dam—An obstruction in a stream or river that holds back the flow of water or raises the water surface elevation so that it may be diverted or pumped into a canal or pipeline.

Downstream Control—Implies downstream water level feedback control.

Downstream Feedback Controller—A controller that uses conditions (typically water level or flow rate errors) downstream from a control structure to decide how to adjust that control structure.

Dynamic Wave Velocity—The velocity of a disturbance on a water surface; also called **Celerity**.

Electronic Filter Level Offset (EL-FLO)—An early downstream control method developed by the U.S. Bureau of Reclamation. The control method used simple proportional-integral control, with a filter to account for pool wave reflection to adjust gate position based on water level errors.

Encoder—Device that measures the change in a desired parameter, such as water depth, in small increments. Many encoders use optical signals. Most encoders read relative change. Reading absolute positions with encoders is more difficult.

Equilibrium—For canal control, when conditions are steady and pool inflow equals outflow.

Error Checking—Procedures for processing raw data to check for spurious, unrealistic values.

Error Signal—The difference between a desired and actual condition (e.g., water levels or flow rates).

External Disturbances—Changes in canal inflow and outflow that are not manipulated by the control system.

Fade Margin—Amount of radio or microwave signal exceeding the minimum required for reliable communication.

Feedback Control—A control method whereby measurement of the desired output (variable with desired set point) is used to adjust input to drive the system to the desired output (e.g., desired water level).

Feedforward Control—A control method whereby measured, predicted, or proposed external disturbances are used to adjust input; the assumption is that these control actions will maintain the output at the desired state (e.g., maintain constant water levels).

Filtering—A procedure for removing noise from the measured output so that controlled inputs do not vary unacceptably; this noise includes sensor noise, water surface waves (e.g., from wind), and unmodeled process behavior (e.g., reflection wave in pools under backwater).

Finite-Time Prediction Horizon—Time over which it is desired to bring the output of the system (e.g., water levels) back to the desired values, or alternately the time period over which output errors are evaluated (i.e., in evaluating controller performance).

Firmware—Low-level microprocessor code (e.g., machine language) that is an integral part of the electronic hardware and is not normally considered software.

Flexibility Factor—A multiplier for the flow-rate capacity of a canal that is based on the desire to allow higher than average demand. This factor usually is based on the probability of satisfying user demand.

Flow Mismatch—The difference between inflow and outflow for a canal pool or canal.

Flow-Rate Control—A control method that maintains a constant or near constant flow rate at a particular point in the canal network, typically through an offtake or check structure.

Flume—In this application, a water measurement device; a physical device for which there is a relationship between upstream water level and discharge.

Freeboard—Height of canal lining and canal banks extended above the canal's normal water surface as a safety measure to protect the canal from overtopping.

Frequency, ω_0—Canal response to cyclic input disturbances is defined in terms of the frequency of these disturbances; the resonance frequencies, ω_r, are those frequencies for which the output (typically water level) fluctuates the most.

Frequency Domain—A way of representing the response of a canal to disturbances and controls so that the frequency response can be examined. This domain is very useful for determining whether proposed controls are stable (i.e., not subject to continuously growing oscillations). See **Bode Plots or Response** and **Nyquist Stability Criterion**.

Frequency Response—The response of a canal to cyclic input disturbances at different frequencies. See **Bode Plots or Response** and **Nyquist Stability Criterion**.

Full Decoupler—A decoupling scheme that keeps disturbances in a given pool from moving either to upstream or downstream pool. See **Decoupler I** and **Decoupler II**.

Full Duplex—Communication channel type in which data flow in both directions simultaneously.

Gain—The ratio of the output to the input signal of a control system; used to describe the proportionality factor of a proportional control system.

Gain Crossover Frequency—Frequency at which the gain is unity.

Gain Margin—Magnitude of the reciprocal of the open-loop transfer function, evaluated at the frequency where the phase angle is equal to −180°.

Gain Matrix—For a controller, the coefficients that determine the next control action (gate flow or position changes) based on the state of the system (e.g., water level errors).

Gain Scheduling—A procedure for adjusting the gain matrix based on the state of the system; for example, changing the controller constants as a function of flow rate.

Gate—A mechanical device that can be incrementally opened or closed to allow water to pass through.

Gate Position Controller—A controller that moves a gate to a desired position; also called **Gate Set Point Controller**.

Gate Position Sensor—A device, such as an analog or digital sensor, that can measure the mechanical position of a gate and provide a signal representing the position.

Gate Set Point Controller—A controller that moves a gate to a set point or desired position; also called **Gate Position Controller**.

Gate Stroking—A mathematical procedure for routing flow changes through a canal without changing water levels at the downstream end of each pool (theoretically).

Gate Trajectories—A set of gate positions as a function of time (time series).

Generalized Predictive Control (GPC)—A control method that determines optimal control trajectories over a finite time horizon based on predicted system response. This controller is optimized each time step based on the current state of the system (i.e., online optimization) but is rather flexible in its ability to handle varied constraints (e.g., limits on gate position or movement).

Global Control—A control system that uses all outputs for the system (e.g., all canal water levels) to determine all control actions (e.g., all changes in gate position or flow); same as **Centralized Control**.

Global Optimization—A method for determining controller constants (tuning) whose criteria considers the response of the entire canal; can be used for either centralized or local controllers.

Gradient-Search Procedures or Gradient-Based Methods—A method for searching for the optimal controller coefficients based on the slope of the response surface, where the response surface represents the performance of the controller as a function of the coefficients.

Gradient Vector—Direction of the tangent (or secant) to the controller response surface.

Half Duplex—Communication channel type in which data flows in both directions, one direction at a time.

Hardware—Physical components of control systems (e.g., relays, switches, lights, integrated circuits, transistors, and capacitors).

Hessian Approximation—An approximation to the **Hessian Matrix**; for example, may be found numerically.

Hessian Matrix—An *n*-by-*n* matrix of second partial derivatives for a function of *n* variables; a convex function (i.e., one with a maximum) requires that the Hessian Matrix be non-negative.

Heuristic—Based on subjective rules, rules of thumb, or empirical data.

Hexadecimal—A number represented in base 16. A through F generally represent numbers 10 through 15.

Hierarchical System—A system with different levels, each higher level composed of and subsuming the lower levels.

High-Frequency Oscillations—In control theory, fluctuations in output that occur over a short time interval (short relative to the response time of the system); often caused by poorly tuned feedback control.

Hydraulic Filter Level Offset (HyFLO) Method—An early canal control method developed by the Bureau of Reclamation based on proportional-integral control concepts with filtering for wave reflection. The method was superseded by EL-FLO.

Hydraulic Interactions—In canal control, influences of flow rate and water levels within one canal pool on the flow rate and water levels in adjacent pools.

Hydraulic Transient—A wave or pressure change propagated through a canal or pipeline during unsteady flow.

Hydraulically Coupled—Canal pools that have hydraulic interactions between pools.

Identification Procedure—A method for determining the value of parameters that define the model of a system; identification can be applied to physical parameters such as Manning's *n*, to model parameters such as pool delay time, or to control parameters such as the proportional constant for a PI controller.

Ill-posed—A problem is called "ill-posed" or "poorly posed" if the solution (output) changes a great deal for only small changes in the input; this is a significant problem for identification of parameters, because measurement noise can have a large influence on the results.

In-Channel Storage—Water storage volume in a canal above the minimum water level required for conveyance.

In-Line Reservoir—A reservoir constructed in line with the canal used to regulate flow for balanced operation.

Incremental Form—In control theory, when control actions are expressed in terms of the change in value from the current value, rather than in terms of absolute value (for example, change in gate position rather than gate opening); also called **Discrete Difference Form** or **Velocity Form**.

Input/Output (I/O)—The input and output signals from an electronic device, in this application, of a computer-based control system or SCADA.

Input/Output Boards (I/O Boards)—An electronic device that sends and receives I/O signals.

Instabilities—In control theory, when output oscillations continue to grow without bounds.

Integral Feedback Control—A feedback control method whereby control actions are taken based on the integral of the deviation of the output from the set point. This usually is used in combination with proportional control and helps eliminate steady-state errors.

Integral of the Absolute Error (IAE)—An integral over time of the absolute difference between the target and actual values of the output; for canal control, usually water level errors.

Integral of the Squared Error—An integral over time of the square of the difference between the target and actual output values (e.g., square of water level errors).

Integrated Absolute Discharge Change (IAQ)—Integral of the absolute changes in discharge at a check structure (controlled flow changes) less the desired change in flow at that structure.

Integrator-Delay (ID) Model—A model of canal pool response that considers a pure delay in the upstream part of a canal pool that is flowing at normal depth and assumes that water level changes are linearly related to the integral over time of net pool inflow.

Interface—A combination of electrical, mechanical, or other components interconnecting elements of a control system.

Inverse Solution—Any of a variety of methods for deducing system inputs or system model parameters from a trace of system outputs.

Inverted Siphon—A conduit through which water flows that is located well below the water surface elevation. Water flows through the conduit because of the difference in water surface elevation between the two sides. For canals, they are commonly used to cross roads and stream channels.

ITRC Flap Gate—A vertical hydraulic gate used to maintain constant upstream water levels automatically; similar designs, such as the

Begemann gate, were developed in the 1940s in Holland, but no design procedure was published.

Jacobian Matrix—A matrix of partial derivatives that indicate how several functions are influenced by common variables.

Kalman Filter—A mathematical procedure for estimating the state of a system from outputs and prior states; this state-estimation procedure functions as a filter, because it conditions the measured outputs prior to determination of control actions.

Kinematic Shock Wave—The steady-state wave produced by a change in discharge in a canal or river; this is the wave shape after all transients have dissipated.

Kinematic Shock Wave Velocity—The steady-state velocity of a wave in a canal or river caused by a change in discharge; this velocity is based solely on continuity and normal depth relationships.

Lagrange Multiplier—In linear programming, coefficients applied to the constraints when the constraints are made part of the objective function.

Lagrangian Formulation—In linear programming, used when the constraints are made part of the objective function.

Langemann Gate—An overshot gate that hinges in the middle to reduce pressure differences usually associated with overshot gates.

Laplace Transform—A method for transforming a system of ordinary linear differential equations into a system of algebraic equations. In control theory, many problems are easier to solve in the Laplace domain rather than the time domain; the Laplace domain also allows one to examine the frequency response of systems.

Lateral—A branch canal or pipeline that diverges from the main canal or other branches.

Leap-Frog Scheme—A method of solving the unsteady equations of open-channel flow by alternately solving for depth and discharge at successive node points.

Limit Switch—Device to prevent overloading of gates and valves by limiting operation or travel to restricted specifications.

Linear Control Methods or Theory—Control methods or theory based on linear systems theory or a set of ordinary linear differential equations.

Linear Displacement Transducer (LDT)—Device used to measure and transmit gate position.

Linear Quadratic Gaussian Regulators (LQG)—Controllers or regulators based on linear systems theory, quadratic performance criteria, and the assumption of Gaussian white noise disturbances.

Linear Quadratic Regulators (LQR)—Controllers or regulators based on linear systems theory and quadratic performance criteria, where the control penalty function is the sum of the squares of the deviations in output (e.g., water level errors), and the sum of the squares of control actions (times some multiplier).

Linearization—Replacement of a nonlinear equation by an approximate linear equation.

Little Man Controller—A mechanical/electrical controller developed to control water levels using a three-position (raise—off—lower) control mode and sometimes an anti-hunt device.

Local Controllers—Controllers that perform a simple function at one location; in canal control, controllers that maintain a water level in an adjacent pool or flow rate through a structure.

Locally Linearized Models—Models of nonlinear systems that are put into linear form based on some initial (or current) conditions.

Long-Crested Weir—A long, overflow manual check structure used in canals to maintain an upstream water level within a reasonable tolerance.

Long-Throated Flume—A water-measuring flume where the length of the throat is roughly twice as long as the upstream head.

Lumped Model—A model of a system based on a correlation of outputs with inputs.

Master Controller—A controller that provides control signals to lower level controllers.

Master Station—The centralized facility with communications to RTUs for the purpose of information retrieval, control of actuators, system control, and operation optimization.

MATLAB—A mathematical software system that contains toolboxes developed for the analysis and design of control systems.

Maximum Absolute Error (MAE)—As used in canal control, the maximum deviation in the absolute value of water level from the set point.

Microwave Radio—A method of point-to-point radio transmission using the frequency spectrum above 890 megahertz (MHz); microwave frequency characteristics limit transmission to line of sight distances.

MIMO (Multiple Inputs, Multiple Outputs)—A system with multiple inputs and multiple outputs.

Minimum Gate Movement—In canal control, the minimum amount of gate movement recommended before the command actually to move the gate is made.

MODBUS—A radio communications protocol for remote terminal units developed by Modicon, Inc.; a de facto standard for radio communications.

Model Inversion—See **Inverse Solution**.

Model Predictive Control (MPC)—A control method that determines optimal control trajectories over a finite time horizon based on predicted system response; this controller is optimized each time step based on the current state of the system (i.e., online optimization) but is rather flexible in its ability to handle varied constraints (e.g., limits on gate position or movement).

Modern Control Techniques—A set of control methods that are based on optimization methods.

Monovariable Controllers—Controllers with one input and one output; also called **SISO**.

Multiplexor—Device that converts multiple measured analog or status signals into a single analog or digital signal.

Multivariable Controllers—Controllers that have multiple inputs and multiple outputs; also called **MIMO**.

Newton Method—Numerical method for solving for the roots of nonlinear equations by using the tangent to the function to determine the next guess; commonly used in optimization problems. It is sometimes shorthand for the "Newton-Raphson" procedure for solving nonlinear equations, which provides adjustments for faster convergence.

Nonconvex Problem—Numerical problem such as minimization problem where either the function to be minimized or the set of variables entering this function are not convex; these problems are difficult because there is no systematic way to prove a minimum solution exists or can be found.

Nonlinear Optimization—Methods for finding the optimum solution for systems that are nonlinear (are not properly represented by linear equations).

Nyquist Plot—See **Polar Plot**.

Nyquist Stability Criterion—A graphical technique for determining the stability of a control system. It is applied to linear, time-invariant systems. It uses a Nyquist plot where the real part of the transfer function is plotted on the X axis and the imaginary part is plotted on the Y axis. A curve shows the open-loop response for each frequency. When this curve does not encircle the point (–1,0) the controlled system is stable in closed loop.

Obermeyer Gate—An overshot gate with an air bladder that is used to adjust the gate position.

Objective Function—The function that is to be minimized or maximized in an optimization problem.

Observability—Indicates whether the state of the system can be estimated by the measured output.

Observer—In control theory, a mathematical procedure used to estimate the state of a system to be controlled through the use of a mathematical model of the system.

Observer Gain Matrix—The matrix used to infer the current state of the system from the prior state of the system and prior control actions.

Observer Poles—In control theory, the poles of the observer or the eigenvalues of the A matrix in a state-space representation.

Off-Line Reservoir—A reservoir constructed to the side of the main canal, usually in a natural drainage channel.

Offtake—A structure that diverts water from an irrigation canal to a lower level canal within the distribution system or a farm delivery point. An offtake from one canal is the headgate for a lower level canal; also known as a **Turnout**.

On-Demand System—An irrigation water distribution system (canal or pipeline) that allows users to extract water without supplying advance information regarding flow rate, time of extraction, or duration of extraction.

On-Line Estimation of the Parameters—Methods for determining parameter values prior to each control action, generally requiring optimization methods.

On-Off Control—See **Two-Position control**.

Open-Loop Control—Control method that adjusts the inputs based on measured disturbances but without measurement of system output.

Operating Criteria—Design and institutional criteria that determine the operating limits of a water system.

Operational Flexibility—For canal delivery systems, the ability to adjust operations to satisfy user demands.

Optimal Controllers—Controllers based on mathematical optimization of performance.

Ordinary Linear Differential Equation—A differential equation with ordinary, as opposed to partial, differential equations where the variable of interest appears once in each additive term.

Oscillations—A cyclic pattern of output, varying over time.

Outlet—A structure that allows water to leave a canal; these include canal spills to drainage systems or offtakes to lower level canals or farms.

Overshooting/Overcorrecting—In control, when too much control adjustment is made and the output goes from one side of the set point to the other, often too quickly and too strongly.

Overshot Gate—A gate that is hinged at the bottom and allows water to flow over the top edge. Generally, the gate leaf is made with one or more flat plates of steel. Flow calibration is similar to a sharp-crested weir.

P+PR—Proportional plus Proportional Reset (P+PR) algorithm is similar to the EL-FLO + Reset algorithm except that the P+PR algorithm is applied to the automatic upstream control of canal structures.

Parameter Estimation—An automatic optimization procedure that determines a set of parameters for a modeled system that provides the closest fit between measured and simulated results, usually evaluated based on quadratic differences.

Parameterization—The process of determining values for a set of parameters or coefficients.

Partial Decoupler (Decoupler I)—See **Decoupler I**.

Penalty Function—In canal control, a function or equation that mathematically describes the negative value or penalty associated with water level deviations, flow rate adjustment, or other things. This function is used to evaluate the quality of a control method or as the objective function in optimization.

Penalty Matrix—A matrix that, in part, defines the penalty function.

Performance Index—An index or numerical value that is used to judge performance (e.g., ability to maintain constant water levels).

Perturbation—Disturbance, noise, and fluctuation.

Perturbation Estimator—A method for determining the value of a parameter associated with a disturbance.

Phase Margin—180° plus the phase angle of the open-loop transfer function at unity gain.

PI Control—See **Proportional-Integral Controller**.

PI+S Control—Proportional-integral control that uses a Smith Predictor to account for delays.

PIR Control—a PI controller developed by Deltour that uses both a Smith Predictor to account for delays and Decoupler I.

Polar Plot—a plot of the Nyquist stability criterion that evaluates the influence of the poles and zeros on the system response; also called **Contour Evaluation** or **Nyquist Plot**.

Pole Placement Methodology—A method for tuning simple controllers based on the solution of the transfer function response equation expressed in the Laplace domain.

Poles—Roots of the denominator of the system response when expressed in the Laplace domain; if these poles are on the right half plane in continuous time or outside the unit circle in discrete time, the system response is infinite (i.e., unstable control).

Potentiometer—A resistance element where a third connection is moved along the resistance element to provide a resistance proportional to the rotation (e.g., a voltage divider).

Power Supply—An electrical device that provides electrical power at a specified voltage in either direct (DC) or alternating (AC) current.

Prediction Horizon—Time period over which a predictive controller evaluates the response of the system or the time to bring the output back to the set point; used in MPC and GPC.

Predictive Control—A control method that uses a finite prediction horizon and online optimization to determine a control strategy.

Pressure Transducer—Device used to convert measured pressure differences to an electrical signal; typically used to measure water depth.

Process Model—A mathematical relationship that describes a physical process.

Programmable Logic Controller (PLC)—An electronic device that accepts analog or digital inputs, makes logical decisions based on these inputs, and sends control signals to external devices.

Proportional Feedback Control—A controller whose adjustment is based on the deviation from the set point.

Proportional-Integral Derivative (PID) Control—A controller whose adjustment is based on the deviation from the set point (proportional), the cumulative deviation from the set point (integral), and the rate at which the output is moving relative to the set point (derivative).

Proportional-Integral (PI) Controller—A controller whose adjustment is based on the deviation from the set point (proportional) and the cumulative deviation from the set point (integral).

Proportional Plus Reset Control—A particular way of implementing PI control, whereby the offset of proportional control is eliminated with the reset action.

Pumping Plant—A system that uses pumps to move water to a higher elevation, usually from a lower to a higher canal. Pumping plants usually include pumps, pipes, electrical panels, electrical switches, and trash screens.

Quadratic Control Criteria—Criteria for the evaluation or development of a controller that are based on the square of the appropriate parameters; for canal control, these are typically the square of the water level error and the square of the required check gate or flow adjustments.

Quadratic Penalty (Objective) Functions—Same as **Quadratic Control Criteria**.

Radial Gate—A gate that is curved in a radius and hinged in the center of this radius so as to reduce the lifting force required to open and close it.

Radio Frequency Interference (RFI) filters—A resistor-capacitor-inductor circuit used to reduce high-frequency transients in electrical circuits.

Random Measurement Noise—Changes in output or input that appear to have no cause and that occur at random.

Rated Section—A cross section of a canal for which a stage (water level) versus discharge relationship has been established.

Reference Scenario—A set of conditions that acts as a reference for comparison. In canal control, the reference scenario could be a constant water level at the set point.

Reflection Waves—Hydrodynamic waves in a canal that result from the reflection of a canal wave off of a structure, an obstruction, or a change in the canal cross section.

Regulation Interval—The time interval between successive control actions for a controller/regulator.

Regulation Period—Same as **Regulation Interval**.

Regulation Reservoir—A reservoir used in canal systems to reduce the mismatch between downstream demands and upstream water supplies to maintain a balanced operation; also known as a **Buffer Reservoir**.

Regulation Time Step—Same as **Regulation Interval**.

Relative Gains—Controller parameters/gains that are expressed in relative or nondimensional form.

Relative Gain Array (RGA[G])—Matrix computed from the transfer matrix G of the system, providing a steady-state measure of interactions for decentralized control.

Remote Monitoring and Control System—A system that is capable of obtaining sensor data from remote sites, assembling them for display, and sending control signals back to these remote sites for action.

Remote Monitoring System—A system that is capable of obtaining sensor data from remote sites and assembling them for display.

Remote Terminal Unit (RTU)—A microprocessor-based unit that reads sensor values (e.g., water levels and gate positions), makes control actions (gate position changes), and communicates with a central station.

Repeatability—The ability of an instrument to read the same value of a data parameter consistently.

Resolution—The smallest distinguishable increment into which a measured quantity is divided. Sometimes expressed in terms of an integer number of "bits," 3 bits $= 2^3 = 8$. This means that the measurement range can be divided into eight increments. If the measurement range of the device is 8 cm, the resolution is 1 cm.

Resonance Frequency—Frequency at which the resonance peak amplitude occurs.

Resonance Peak Amplitude—Maximum value of the magnitude of the closed-loop frequency response.

Response Time—The time required for the depth, pressure, or flow to reach and remain within a certain percentage of its steady-state value after a control correction has been initiated.

Riccati Equation—An equation that expresses the state transition equations and penalty functions for a linear quadratic regulator in a way that allows a relatively straightforward solution for the optimal gain matrix.

Right Half Plane Zeros (RHPZ)—These result from positive roots in the numerator of the Laplace transfer function. An RHP Zero imposes limitations on the performance that can be obtained when controlling this nonminimum phase system.

Rise Time—Time interval for the response to change from 10% to 90% of its steady-state value; sometimes other percentages are used.

Robustly Stable Control System—A control system that maintains stable control over a wide range of conditions.

Robustness—Ability to function over a wide range of conditions.

Rotation Delivery—Water delivery whereby a single, relatively constant supply flow is rotated to different users at varying times, with any one user receiving 100% of that flow rate at a time.

Saint Venant Equations—Hyperbolic, nonlinear, first order, partial differential equations that describe continuity and momentum for unsteady water flow (also called De Saint-Venant Equations).

Saturation of Inputs—A condition that occurs when a controlled input to a system cannot be changed further (e.g., in canal control when a gate is already closed or already open out of the water).

SCADA Integrator—An individual or company responsible for installing (and often selecting) all or most of the components (software and hardware) required for a SCADA system, both in the field and in the office.

Scheduled Delivery—Operation of a water delivery system to meet predetermined needs, generally based on user water orders; also called **Arranged Delivery**.

Self-Regulation—A controlled system requiring virtually no operation intervention. See **Automatic Control**.

Self-Tuning Control—A controller that adjusts automatically its control parameters to remain tuned.

Sensor—A device for measuring water level, flow, gate position, and other parameters for input to a local automatic controller or remote terminal unit (RTU) or for monitoring.

Set Point—The desired value of process output (or parameter that is to be held constant); in canal control, usually the desired water level.

Service Level, 90%—Usually refers to the capacity of a canal or pipeline that will be able to meet the requested demand 90% of the time.

Settling Time—Time interval between the step change and the time for the response to be within the allowable tolerance.

Shallow Water Equations—Unsteady equations of continuity and momentum used for simulating canal transients. See **Saint Venant Equations**.

Sharp-Crested Weir—A water measurement weir constructed from a thin plate of steel; the top edge of the plate is beveled to provide a more accurate rating.

Signal Conditioning—Method of improving the quality of an electronic signal over distance.

Simplex—Communication type in which data flow in only one direction.

SISO (Single Input, Single Output) Controller—A controller (or system) with a single output of interest and a single controllable input.

Slave Controller—A controller that receives signals (e.g., set point information) from a higher level, more global (master) controller.

Sluice Gate—A vertical gate that is raised and lowered to allow water to pass under and/or around it.

Smith Predictor—A method applied to feedback control to account for long time delays between control actions and system output response; generally uses a linear process model to predict the effect of prior control actions on system output.

Spread-Spectrum Radios—Radios that avoid noisy frequencies by automatically switching between many frequencies to transmit data.

Stability—The state of being stable.

Stable—A system where disturbances do not cause the system output to grow without bounds.

State Cost Weighting Matrix—A matrix that contains the relative weights for parameters that describe the system. These parameters have a cost or negative value associated with not having a particular value (typically zero). In canal control, these would represent the relative value of various water levels deviating from their set points.

State Estimator—A mathematical method for computing values for parameters that describe the state of the system; the **Kalman Filter** is a common state estimator.

State Feedback—A feedback control system or method based on state-transition relationships.

State Observer—Same as **State Estimator**.

State-Space Equation—An equation that describes the relationship among system states, inputs (control actions and disturbances), and outputs (controlled variables and measured variables) as a function of discrete time.

State-Transition Equations—An equation that describes the future (next time step) state of a system in terms of prior states and prior inputs (control actions and disturbances) in discrete time.

State-Transition Matrices—Matrices whose parameters describe how the system state changes as a function of current and prior states and inputs.

State Variables—Parameters that define the state of a system; these can be physical variables or mathematically determined variables, which are related to the physical variables.

State Vector—A vector that contains the various state variables that describe a system.

Static Error—Same as **Steady-State Error**.

Steady-State Error (StE)—An error in system output that is not corrected, even after a long time.

Stilling Well—A cavity, generally a vertical pipe in the bank of a canal, where the water level in the canal is measured; waves in the canal are damped ("stilled") by a pipe or opening connecting the canal to the stilling well.

Storage Reservoir—A location where water can be stored for subsequent use. It can be natural or man-made. Often results from the blocking of a natural waterway with a dam. Can be constructed by digging a pit or by constructing a series of berms to contain the water.

Supervisory Control—The control of a canal system from a centralized location (master station) over a communication system; uses RTUs at the canal structure sites.

Supervisory Control and Data Acquisition (SCADA)—A centralized control system that retrieves information from remote sites, processes this data for display, stores the data for future use, and sends control signals or new targets to remote sites.

Supply Oriented—A canal control system that takes the water supply and distributes it to users, often without sufficient consideration of the need for water; in opposition to "demand-oriented."

Surge Suppression—A method for removing unwanted and potentially dangerous signals (i.e., surges) from influencing the behavior of a system; this is applicable to surges of electrical power to computer equipment, surges of pressure in a pipeline, surges of flow/level in a canal, and surges in disturbances for a control system.

System Dynamic Equation—An equation that describes the dynamic behavior of a system.

System Identification—Same as **Parameter Estimation**, although system identification also can determine which parameters are significant.

Systems Integration—The combining of different, often incompatible components into a working system.

Target Values—Same as **Set Point**.

Telemeter—To sense, encode, and transmit data to a distant point.

Three-Position Control—A mode of control that responds to the deviation from the set point by operating the controlled device for a predetermined amount of time. The corrective signal has three discrete values: no correction, increase the output correction, and decrease the output correction.

Time Lag—In canal control, the time between a control action (e.g., change in gate position) to the associated change in output (e.g., water level response).

Tracking—Regulation when the set point is a varying function of time.

Transducer—A device that converts one form of energy to another such as hydraulic to pneumatic or mechanical to electrical.

Transfer Functions—Mathematical expressions that define the relationships between various inputs and outputs of a system.

Translatory Wave—A gravity wave that propagates in an open channel and results in displacement of water particles in a direction parallel to the flow.

Transmission zeros—See **Zeros**.

Tuning—A process of determining the parameters or gain matrix for a controller.

Turnout—A structure that diverts water from an irrigation canal to a lower level canal within the distribution system or a farm delivery point; a turnout from one canal is the head gate for a lower level canal; also known as an **Offtake**.

Two-Position Control—A mode of control that responds to the deviation from set point by operating the controlled device to either of two extreme positions; also referred to as **On-Off Control**.

Ultra High Frequency (UHF)—Ultra-high frequency type of radio system (300 to 3,000 megahertz).

Uninterrupted Power Supply (UPS)—An electronic device that regulates electrical power output and provides power for short periods of time even when the supply voltage is well below normal, usually through the use of batteries.

Unstable—A system whose output grows without bound; control systems that have significant oscillations are often considered to be unstable if the oscillations exceed some threshold, even if they do not continue to grow.

Unsteady Flow—Flow that is changing with respect to time.

Upstream Control—Implies upstream water level feedback control. (1) A means of control intended to maintain a target water level or pressure upstream from a single gate or valve; the structure may be automated or manual. (2) A control concept whereby the first in a series of structures delivers a flow rate based on some water master controlled schedule while the remaining structures are adjusted to control the water level immediately upstream.

Variable Frequency Drive (VFD)—An electronic control technique that varies the input frequency (Hz) of the alternating current that is supplied to a motor, enabling the motor to be operated at a range of speeds; typically used to control pump discharge.

Velocity Form—A form for control equations, whereby new controller positions (control actions) are expressed relative to the existing positions (i.e., changes in positions), and system outputs are expressed relative to set points.

Very High Frequency (VHF)—Type of radio system (30 to 300 megahertz).

Volume Compensation—A method of canal control that considers the volume changes explicitly that occur in a canal as a function of flow rate, set point, and other parameters.

Wasteway—A structure or channel that is used to dispose of excess water.

Water Master—The person responsible for operation of the entire canal project.

Wave Celerity—The velocity of propagation of a wave through a liquid, relative to the rate of movement of the liquid through which the disturbance is propagated.

Wedge Storage—The volume of water contained between two different water surface profiles within a canal pool.

Weir—In this context, a water measurement device that relates upstream water level to discharge.

Zeros—Roots of the numerator of the system response equation when expressed in the Laplace (or Z) domain.

Ziegler-Nichols Tuning Method—A method for tuning simple, classical controllers based on measured system response and rules of thumb; may not be useful for multiple, interacting systems, such as multipool canals.

INDEX

Page numbers followed by *e*, *f*, and *t* indicate equations, figures, and tables, respectively.